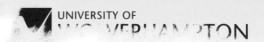

FOOD MICROBIOLOGY

FOOD MICROBIOLOGY
A Laboratory Manual

AHMED E. YOUSEF
CAROLYN CARLSTROM

The Ohio State University

A JOHN WILEY & SONS, INC., PUBLICATION

Copyright © 2003 by John Wiley & Sons, Inc. All rights reserved.

Published by John Wiley & Sons, Inc., Hoboken, New Jersey.
Published simultaneously in Canada.

No part of this publication may be reproduced, stored in a retrieval system, or transmitted in any form or by any means, electronic, mechanical, photocopying, recording, scanning, or otherwise, except as permitted under Section 107 or 108 of the 1976 United States Copyright Act, without either the prior written permission of the Publisher, or authorization through payment of the appropriate per-copy fee to the Copyright Clearance Center, Inc., 222 Rosewood Drive, Danvers, MA 01923, 978-750-8400, fax 978-750-4470, or on the web at www.copyright.com. Requests to the Publisher for permission should be addressed to the Permissions Department, John Wiley & Sons, Inc., 111 River Street, Hoboken, NJ 07030, (201) 748-6011, fax (201) 748-6008, e-mail: permreq@wiley.com.

Limit of Liability/Disclaimer of Warranty: While the publisher and author have used their best efforts in preparing this book, they make no representations or warranties with respect to the accuracy or completeness of the contents of this book and specifically disclaim any implied warranties of merchantability or fitness for a particular purpose. No warranty may be created or extended by sales representatives or written sales materials. The advice and strategies contained herein may not be suitable for your situation. You should consult with a professional where appropriate. Neither the publisher nor author shall be liable for any loss of profit or any other commercial damages, including but not limited to special, incidental, consequential, or other damages.

For general information on our other products and services please contact our Customer Care Department within the U.S. at 877-762-2974, outside the U.S. at 317-572-3993 or fax 317-572-4002.

Wiley also publishes its books in a variety of electronic formats. Some content that appears in print, however, may not be available in electronic format.

Library of Congress Cataloging-in-Publication Data:

Yousef, Ahmed Elmeleigy.
    Food microbiology : a laboratory manual / Ahmed E. Yousef, Carolyn Carlstrom.
    p. cm.
    "A Wiley-Interscience publication."
    Includes bibliographical references and index.
    ISBN 13: 978-0-471-39105-0
    1. Food—Microbiology–Laboratory manuals.   I. Carlstrom, Carolyn.   II. Title.

QR115 .Y686 2003
664'.001'579—dc21

2002032426

CONTENTS

PREFACE — vii

PART I. BASICS OF FOOD MICROBIOLOGY LABORATORY — 1

  1. BASIC MICROBIOLOGICAL TECHNIQUES — 5

PART II. FOOD MICROBIOTA — 23

  2. TOTAL PLATE COUNT — 27
  3. YEASTS AND MOLDS — 42
  4. COLIFORM COUNT IN FOOD — 61
  5. MESOPHILIC AEROBIC AND ANAEROBIC SPORES — 81
  6. MICROBIOTA OF FOOD PROCESSING ENVIRONMENT — 97

PART III. FOOD-TRANSMITTED PATHOGENS — 111

  7. *Staphylococcus aureus* — 121
  8. *Listeria monocytogenes* — 138
  9. *Salmonella* — 167
  10. *Escherichia coli* O157:H7 — 206

PART IV. FOOD FERMENTATION   223

11. LACTIC ACID FERMENTATION AND BACTERIOCIN PRODUCTION   231

APPENDIX A   LABORATORY EXERCISE REPORT   249

APPENDIX B   MICROBIAL GROWTH KINETICS   257

APPENDIX C   MICROBIOLOGICAL MEDIA   261

INDEX   275

PREFACE

A typical course in food microbiology provides a basic education relevant to spoilage, pathogenic and beneficial microorganisms. The practical component of such a course should train students in methods to enumerate spoilage and indicator microorganisms, detect pathogens most common in food, and assess the beneficial aspects of useful microorganisms. Our food microbiology laboratory manual was developed with this in mind. The manual has four main parts. Part I reviews basic microbiological techniques which may have been covered in previous elementary biology or microbiology classes. Part II includes exercises to evaluate the microbiota of various foods and enumerate indicator microorganisms. Conventional cultural techniques are emphasized in the first two parts. The goal of Part III is to provide practice in methods of detecting pathogens in food. During this part, students will have chances to practice cultural, biochemical, immunoassay, and genetic methods. Finally, Part IV presents beneficial microorganisms and their role in food fermentations. Production of food preservatives, namely lactic acid and bacteriocins, through a batch fermentation will be specifically assayed for.

During the past 10 years, the contents of this laboratory manual have been developed and exhaustively tested at The Ohio State University (OSU), Departments of Microbiology and Food Science and Technology. Most of the laboratory exercises were run repeatedly and successfully by undergraduate and graduate students taking the OSU food microbiology course. The manual is suitable for a laboratory component of a food microbiology course or as a standalone laboratory course. Students taking this course are expected to have a suitable background in basic microbiology.

Methods in this manual are meant to train students to assess the microbiological quality and safety of food. We want to emphasize that the exercises in this manual are not intended to substitute or even replicate approved standard or official methods for microbiological analysis of food. Methods included were customized to fit two-hour laboratory periods and for a class that meets two or three times a week for 12–14 weeks. Although the manual was carefully reviewed, errors may still be discovered. We appreciate you sharing these errors with the authors so that improved versions of the manual can be developed in the future. Clarification of material from the book can be obtained from a companion web site: http://class.fst.ohio-state.edu/fst636/fst636.htm. Color illustrations of results obtained when running exercises and examples of laboratory presentations can be found on the website.

Finally, we would like to acknowledge the help we received in the past 10 years from all the graduate students who participated in the laboratory instructions that led to developing and fine tuning the experiments in this manual. We thank Maya Achen, Tessa Blais, Julia Byard, Hyun-Jung Chung, Sandhya Dave, Julie Jenkins, Jin-Gab Kim, Xia Liu, Yuqian Lou, Bindhu Michael, Laura Reina, Luis Rodriguez-Romo, Ragip Unal, Mary Tichi, Wendy Tsai, and Amy Wanken for participating in laboratory instructions. We particularly thank Dr. Yoon-Kyung Chung for her help in developing the yeasts and molds exercise, Kristen Anderson and Adraine Liner for developing and writing the early draft of the *Escherichia coli* O157:H7 exercise, and Beatrice Lado for fine tuning the example laboratory report and worksheet in Appendix A. Dr. Chung was also invaluable in reviewing the final manuscript. We also thank the hundreds of students at OSU who practiced these methods over the past 10 years and pushed us to provide a better manual.

We recognize the fine contributions to this project by Lisa Robinson, the instructional assistant, Department of Microbiology, OSU. She proofread the exercises and helped us troubleshoot the problems encountered when we developed new procedures. It was always a pleasure working with Lisa and seeing her excited about trying new protocols. We would like to thank Don Ordaz and Jon-David Sears for their help in setting up and trouble shooting the automation required for the fermentation exercise. We wish to acknowledge the OSU Microbiology Department's Laboratory Coordinator, Elizabeth Wrobel-Boerner, whose example in experiment planning led us to higher expectations for what students could accomplish. Posthumous appreciation goes to Dr. Stuart Wahl, who devoted numerous hours to improving the laboratory exercises while he was participating in the laboratory instructions. We appreciate Dr. Olli Tuovinen's support while Carolyn Carlstrom spent time away from his laboratory to work on this project.

Ahmed E. Yousef (yousef.1@osu.edu)
Carolyn Carlstrom (carlstrom.3@osu.edu)

The Ohio State University

PART I

BASICS OF FOOD MICROBIOLOGY LABORATORY

It is essential that one become familiar with the food microbiology laboratory setup and safety guidelines before running any exercises. This not only ensures the safety of workers but also leads to efficient use of their time and laboratory resources. Once this is accomplished, students can review and practice basic techniques covered in previous introductory microbiology courses. Although experienced students may find this redundant with what they learned in other courses, it allows instructors and students to agree on common rules that will be applied in the laboratory exercises.

LABORATORY ENVIRONMENT AND PERSONAL SAFETY

There are many facets to the laboratory environment, ranging from tangible items such as equipment and materials to more conceptual aspects such as safety. Laboratory instructors and students should be familiar with the required equipment and its operation. They should also be familiar enough with the environment to respond appropriately to safety issues and emergencies.

Microbiology laboratory equipment includes everything from basic items such as incubators and refrigerators, water baths, autoclaves, and centrifuges to more modern innovations such as gel electrophoresis systems, multiple-well (or microtiter) plate readers, and polymerase chain reaction (PCR) thermocyclers. Smaller scale equipment may consist of Bunsen burners, streaking loops, thermometers, microscopes, and slides.

To adequately complete the assignments, it is necessary for the laboratory instructor or supervisor to be sure all required equipment is available and in working order. To this end, a list of required equipment is included with each chapter. The student should be sure to understand what the equipment does. For example, the multiple-well plate reader is a spectrophotometer that measures the amount of light absorbed through the well contents, allowing the researcher to determine the amount of substrate reacting in an immunoassay reaction mixture.

Food microbiology involves the study of organisms that are known to cause disease in humans. Therefore, careful work habits are important to prevent the spread of disease to laboratory workers, those they may come in contact with, or others who may use the laboratory space. Familiarity with the laboratory environment itself and the procedures required to keep that environment safe and clean is a key component to good microbiology laboratory work.

Personal Safety

- *Never begin work in the laboratory without the prior permission of the laboratory instructor or supervisor.*
- *Wear a laboratory coat.* Wearing a laboratory coat in the laboratory provides some protection. The laboratory coat may be stored in a designated location in the laboratory.
- *Wash your hands.* Washing hands prevents the worker from contaminating the food sample and from the food sample contaminating the worker. In some cases the use of disposable gloves is recommended or required.
- *Don't eat or drink in the laboratory.* The laboratory environment is not an appropriate place for eating and drinking. In fact, any activity that might involve putting something into the mouth (e.g., chewing gum, applying lip gloss/chap-stick, smoking, or chewing on a pencil) may provide an opportunity for a pathogen to infect the worker. Of course, mouth pipetting is both a poor technique and a safety hazard.
- *As much as possible, keep the bench free of materials except those actually being worked with.*
- *Be familiar with the available safety equipment and supplies.* This includes the locations of the first-aid kit, safety showers, eye wash stations, fire extinguishers, fire blankets, and fire alarms.
- *Avoid fire hazard.* Hair must be pulled back to prevent the possibility of it falling in the burner. Similarly, hats with brims should be avoided as the brim might come near the flame. Alcohol fires are among the most common laboratory fires. Should a jar of alcohol catch fire, placing the lid over the jar will quickly suffocate the fire. Keep flame away from staining bottles as these often contain alcohol. Students should be aware of the location of available fire safety equipment and the nearest exits in case of larger fires. Remember to never use elevators in case of fire. Any fire should be reported immediately to the laboratory supervisor.
- *Report any personal injuries to the laboratory instructor/supervisor.*
- *Follow the guidelines set by the U. S. Centers for Disease Control and Prevention when handling pathogenic live cultures* (CDC/NIH, 1999).

Materials

Typically, each student is assigned a locker or cabinet containing materials commonly used in the laboratory. Students should be sure that the locker contains all of the materials indicated in the course handouts and that all materials are returned to the cabinet at the end of each session. Typical equipment contained in a locker might include an inoculating loop, an inoculating needle, microscope slides, coverslips, microscope lens-cleaner, lens oil, lens cleaning paper, wax marking pencils, pipette bulbs, bibulous paper, and matches. Some of these materials may be used up during the course of the term, and students should learn where replacement materials are kept.

At the beginning of every laboratory session, the student should determine the location of all water baths, incubators, or other equipment that will be shared during that session. Students should collect all required media and supplies. Caution should be taken whenever obtaining media to carefully label them as many media look similar. Do not collect any more media than will actually be used by the group during the exercise. Careful reading of the manual will allow students to determine the correct number of plates and tubes for each exercise.

Materials Handling

A microbiology laboratory contains many materials that are potentially dangerous outside the laboratory environment. Students should never remove slides, plates, or tubes from the laboratory. After use in the laboratory, materials are either prepared for reuse or discarded. Reusable materials include some glassware such as tubes, bottles, and flasks. All labels should be removed from these items, by either peeling off labeling tapes or erasing any writing. This reusable glassware should then be placed in the designated locations for each type of item. Depending on their contents, tubes, bottles, and flasks may be autoclaved before washing. Some items, such as blender jars, may be washed by students. These items should be washed according to the designated protocol and placed in the designated drying area.

Nonreusable materials are disposed of as either hazardous or nonhazardous waste. Paper towels used with disinfectant to wipe off laboratory benches may be placed in containers for nonhazardous waste (i.e., regular trash). Gauze or lens paper used to clean microscope lenses, before or after use, is also safe to place in the regular trash.

Biohazard containers (e.g., specially marked bags in boxes) are used to discard culture-containing disposable Petri plates or tubes so they may be autoclaved or incinerated. Plastic biohazard receptacles (*sharps containers*) are used for sharp items such as pipettes, toothpicks, and microscope slides so that the receptacle may be safely handled prior to disposal. Contaminated broken glassware should be disposed of in the sharps container. Broken uncontaminated glassware should be placed in a receptacle designated for that purpose.

Spilling of cultures can happen. In this case, the worker needs to flood the area with a disinfectant and wipe the area with paper towels or other provided absorbent towels, wiping toward the center to prevent spread of the contaminant. These towels are considered contaminated and should be disposed of in the biohazard container. Appropriate disposable gloves should be worn during this cleaning process.

Students should tape their plates together at the end of each exercise to make retrieving the plates easier at the beginning of the next session. Plates should be placed in the correct orientation and in the designated location.

REFERENCE

Centers for Disease Control and Prevention and National Institutes of Health (CDC/NIH). 1999. *Biosafety in Microbiological and Biomedical Laboratories*, 4th ed. U.S. Government Printing Office, Washington, DC. Available: http://www.cdc.gov/od/ohs/biosfty/bmbl4/bmbl4toc.htm.

CHAPTER 1

BASIC MICROBIOLOGICAL TECHNIQUES

INTRODUCTION

Food processors are increasingly dependent on microbiological analyses to judge the quality and safety of food. A meaningful microbiological analysis is an outcome of properly chosen and executed techniques. Familiarity with preparation, handling, and use and choices of culture media is an important part of any microbiological analysis. Appropriate media contain nutrients to support growth of targeted microorganisms and agents that aid in selection or differentiation of these microorganisms. Development of highly specific media enables microbiologists to make significant inferences about the biochemical and physiological nature of the organisms investigated.

The medium used for microbiological analysis should be sterile to ensure correct interpretation of results. Autoclaving and occasionally microfiltration are used to achieve sterility. Care should be exercised during handling of sterile media to avoid introduction of contaminants; that is, aseptic technique should be followed. Using aseptic technique ensures that only microorganisms of interest, whether they are a part of the food or in pure cultures, are introduced (inoculated) in the medium. Inoculation of agar media includes streaking, pour-plating, and spread-plating methods. Inoculation methods may affect interpretation of results; only qualitative information is obtained using a streak plate, but the pour- and spread-plate methods allow enumeration of organisms (quantitative information).

MICROBIOLOGICAL MEDIA

Culture media are commonly described as nonselective, selective, differential, or selective–differential (Fig. 1.1). Nonselective media are nutritionally rich and com-

---

Food Microbiology By Ahmed E. Yousef and Carolyn Carlstrom
ISBN 0-471-39105-0 Copyright © 2003 by John Wiley & Sons, Inc.

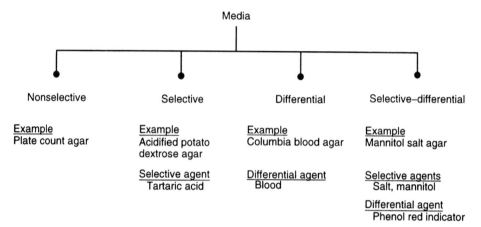

Fig. 1.1. Types of culture media commonly used in microbiological analysis of food.

monly used to enumerate total microflora or to transfer and maintain the purified microorganisms. A selective medium allows the microorganism of interest (targeted microorganism) to grow and suppresses the growth of competing microorganisms. Appropriate selective agents (e.g., crystal violet, which selects against gram-positive bacteria) are essential components of this class of media. Differential media are occasionally used to distinguish between subpopulations in food based on a single biochemical property. Proteolytic and nonproteolytic, lipolytic and nonlipolytic, and acid- and non-acid-producing bacteria are detected and enumerated using simple differential media containing appropriate differential agents. A medium containing both selective and differential agents provides an efficient means of selection and differentiation of the microorganism of interest when present in a complex microbiota. Culture media are used in liquid (broth) or solid (agar) forms depending on the purpose of the experiment.

Media Preparation Guidelines

- Always read the label on the container of the medium before use. Verify and record the medium, the manufacturer, and the expiration date.
- Accurately weigh or measure all ingredients and add them according to the instructions for the procedure being used.
- Use distilled or deionized water.
- If the medium contains agar, stir continuously and heat to boiling before dispensing into bottles. Agar melts (as evident by medium clearing) in boiling aqueous solution. Agar is generally used at 15 g/l, although this may vary depending on the medium and use.
- Any medium requiring sterilization should be placed in a heat-resistant container. Never fill a container more than three-fourths full as the liquid will expand during heating. Screw the caps on loosely to allow pressure to escape. In general, the cap should be a half turn or more from tight. Mark racks

of tubes or media bottles with indicating tape (autoclave tape) to verify sterilization.
- Since agar sets (turns into a gel) at about 45°C, make sure that autoclaved (or steamed) agar media are kept in a waterbath set at about 50°C, until pouring into plates.
- Sterilization is achieved by holding a material or medium for 15 min at 121°C in the autoclave.
- Large loads of material may require more autoclaving time to ensure all items are completely sterilized.
- Use the proper exhaust for the materials being autoclaved. Liquids should be exhausted more slowly than glassware so that the change in pressure does not suck liquid out of the container.

DILUTION AND PLATING

Foods vary in microbial load depending on how they are produced, processed, transported, and handled. Before most food samples are analyzed microbiologically, they should be diluted so that enumeration can be accomplished with reasonable accuracy. A decimal dilution of the food sample is prepared by mixing one part of the sample with nine parts of a diluent, which is commonly a physiological saline or a peptone water. The dilution factor for such a sample is 1/10, or 10^{-1}, and is calculated as follows:

$$\text{Dilution factor} = \frac{\text{sample volume}}{\text{sample volume} + \text{diluent volume}} \quad (1.1)$$

Additional dilutions are prepared as needed and portions of these dilutions are plated (Fig. 1.2). "Plating" refers to the process of transferring and incorporating the sample to be analyzed, or its dilutions, into a suitable agar medium in a Petri plate. When the agar medium is poured and solidified in the Petri plate in advance, incorporation of the sample is done by spreading and the process is described as "spread plating." However, the sample, or its dilution, may be dispensed first in an empty Petri plate, warm molten agar is added, and plate contents are mixed. This process is known as "pour plating."

INCUBATION

Microorganisms vary in their ability to grow at different temperatures. While psychrophiles prefer refrigeration temperatures (1–10°C), mesophiles grow optimally at temperatures close to that of the human body (37°C), and thermophiles grow best at higher temperatures (e.g., 55°C). Psychrotrophic bacteria grow optimally in the mesophilic range, but they also are capable of growing under refrigeration. Choice of incubation temperature, therefore, depends on the type of microorganism and its natural habitat.

8 BASIC MICROBIOLOGICAL TECHNIQUES

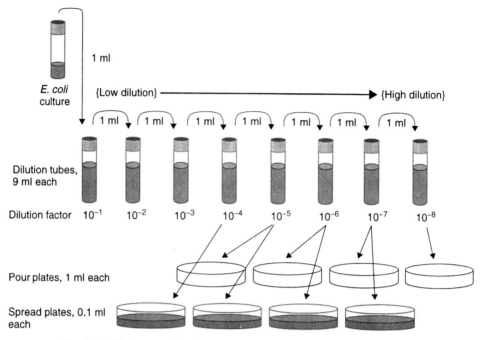

Fig. 1.2. Dilution and plating scheme used in basic technique exercise.

In food microbiology, several incubation temperatures are typically used. Each temperature is used to help enhance the growth of particular types of organisms. For incubation of potentially pathogenic organisms, such as *Salmonella* spp., incubation is done at a temperature similar to that of the human body. Most detection and enumeration methods state a mesophilic incubation temperature of 35°C, although a range of 32–37°C should give acceptable results. A somewhat cooler temperature (e.g., 30°C) is more preferred by yeasts, molds, and psychrotrophic bacteria. "Room temperature" is typically taken to mean 22°C, but this temperature may vary depending on the room used for incubation and even the season and area of the world. Refrigeration at 4°C is typically used to maintain cultures without allowing additional growth. This refrigerator temperature is also used for cold enrichments of psychrotrophic microorganisms such as *Listeria monocytogenes*. Throughout this manual the temperatures 35, 30, 22, and 4°C will be used.

Plates containing inoculated agar media are inverted before incubation. If plates are incubated with lids upward, water condensate falls on the agar surface causing the spreading of colonies. Therefore, plates are inverted during incubation. In this case, condensed water (from moist agar) accumulates on the plate lid. Excessive water condensation on the lid, however, is undesirable and should always be minimized. Pouring hot agar (>50°C) aggravates this problem. In case of spread plates, it is preferable to pour the agar in these plates 24–48 hr before use. Some microbiologists "dry" the spread plates soon after preparation for several hours in a warm, clean incubator. Plate lids with excessive water condensation should be replaced with dry sterile lids.

Fig. 1.3. Dark-field Quebec colony-counter with Petri plate mounted for colony counting.

COUNTING MICROBIAL POPULATIONS IN FOOD

"Counting" in food microbiology refers to determining the number of colony-forming units (CFU), or viable microbial cells, present in a unit volume or weight of the sample. Enumerating the number of colonies on agar plates may also be referred to as "counting"; therefore, careful distinction between these two usages is urged. Some enumeration techniques, such as the microscope count method, allow determining the number of *cells* per unit volume or weight of the sample. The plate count technique, used in this exercise, however, determines the number of cells or cell clumps that are capable of forming colonies on agar plates. Since it is impossible to distinguish colonies arising from individual cells and those from cell clumps, the final population count determined by this method is expressed in colony-forming units per unit volume or weight (i.e., CFU/ml or CFU/g).

Counting colonies on plates can be done visually, preferably with the help of a colony counter. Colony counters, such as the dark-field Quebec colony counter (Fig. 1.3), provide background lighting and magnification so that small colonies are not missed. To carry out this process accurately and rapidly, the analyst should mark the counted colonies, on the bottom of the plate, with a marker pen to make sure that the same colonies are not counted repeatedly.

Knowing the dilution factor (Eq. 1.1), volume plated, and number of colonies on the plate (or average from duplicate plates), count of microorganisms in food is calculated using the following equation:

$$\text{Count(CFU/ml or CFU/g)} = \frac{\text{average number of colonies from duplicate plates}}{\text{dilution factor} \times \text{volume plated}} \quad (1.2)$$

For ease of calculation, 1 ml of diluted food sample is considered equivalent to 1 g.

10 BASIC MICROBIOLOGICAL TECHNIQUES

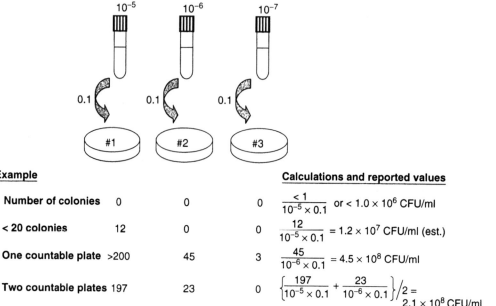

Example				Calculations and reported values
Number of colonies	0	0	0	$\frac{<1}{10^{-5} \times 0.1}$ or $< 1.0 \times 10^6$ CFU/ml
< 20 colonies	12	0	0	$\frac{12}{10^{-5} \times 0.1} = 1.2 \times 10^7$ CFU/ml (est.)
One countable plate	>200	45	3	$\frac{45}{10^{-6} \times 0.1} = 4.5 \times 10^8$ CFU/ml
Two countable plates	197	23	0	$\left\{\frac{197}{10^{-5} \times 0.1} + \frac{23}{10^{-6} \times 0.1}\right\}/2 = 2.1 \times 10^8$ CFU/ml
All > 200	>200	>200	>200	See Fig. 1.5 for estimating this count.

Remember to:
- Include "estimated," abbreviated as "est.," when appropriate.
- Use scientific notation: for instance, 1.0×10^3, not 1000.
- Include only one number past decimal, e.g., 1.7×10^8, not 1.713×10^8.
- Always include units.

Fig. 1.4. Applying colony-counting rules.

Counting Rules

To obtain counts that can be compared among different laboratories, it is necessary to establish consistent guidelines for counting colonies. In some circumstances, however, different counting or calculating methods may be used in place of or in conjunction with the standard counting rules. The following are the counting rules that will be applied throughout this manual, and Figs. 1.4 and 1.5 show samples of calculations using these rules:

1. *Plates with Colonies in the Range of 20–200 (Best Possible Scenario)* If plating yielded plates with colony counts in the range of 20–200 per plate (as judged by a preliminary estimation), discard all remaining plates and count only the plates in this range. Notice that other references use 30–300 colonies or other ranges as suitable countable colonies on a plate. Calculate the CFU/ml using Eq. 1.2. Examples of dilution factors are 1/10 and 1/100 and the volume plated is commonly 1 or 0.1 ml. The numerator of Eq. 1.2 may vary as indicated in the following cases:

   A. ONE PLATE IN 20–200 RANGE If the exercise yields only one plate with a colony count in the range of 20–200, calculate the CFU/ml or CFU/g in the original sample using the number of colonies on that plate instead of an average (Eq. 1.2).

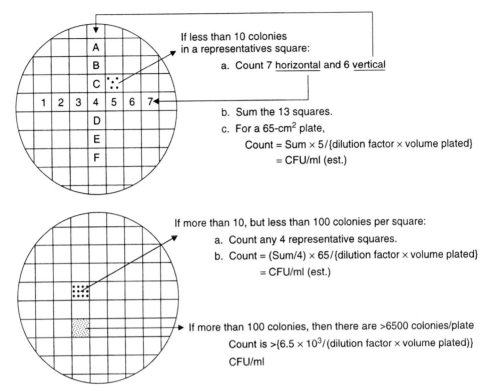

Fig. 1.5. Counting colonies in crowded plates.

B. CONSECUTIVE DILUTIONS If plates from two consecutive dilutions yield 20–200 colonies, compute the CFU/ml or CFU/g resulting for each of the two dilutions. If the numbers are not appreciably different (e.g., 1.5×10^4 and 1.2×10^4 CFU/ml), average the numbers and report the average. If the numbers are substantially different (e.g., the higher CFU/ml is twice the lower one), report only the lower computed CFU/ml or CFU/g.

2. *Plate with 1–20 Colonies* If the plating procedure results in only plates with less than 20 colonies, record the actual number of colonies on the plate receiving the lowest dilution (i.e., the highest concentration plated) and apply Eq. 1.2. In addition, report the number as "estimated" or "est." For example, a sample of cooked meat was analyzed by pour plating 1 ml of sample dilutions 10^{-2}, 10^{-3}, and 10^{-4}. Incubated plates produced less than 20 colonies, and thus the count of microbial population in meat is calculated and reported as shown in Table 1.1.

3. *Plates with no Colonies* When plates produce no colonies, the count is estimated to be smaller than the minimum detection limit of the procedure followed. The minimum detection limit is the count that would result from the presence of one colony on the plate receiving the lowest dilution of the sample. Apply Eq. 1.2 to estimate the count using the lowest dilution plated and substitute the numerator of the equation with <1. For example, if no colonies appeared on any of the plates

TABLE 1.1. Total Plate Count[a] for Cooked Meat

Dilution Factor	Number of Colonies	
	Plate 1	Plate 2
10^{-2} (least dilute)	12	16
10^{-3}	3	1
10^{-4} (most dilute)	0	0

$$\text{Count}^a = \frac{(12+16)/2}{10^{-2} \times 1} = 1.4 \times 10^3 \, \text{CFU/g (est.)}$$

receiving the dilutions shown in the previous example (Table 1.1), then

$$\text{Count(CFU/g)} = \frac{<1}{10^{-2} \times 1} = <1.0 \times 10^2 \, \text{CFU/g}$$

4. *Plate with Greater Than 200 Colonies* If the plating procedure yields only plates with greater than 200 colonies, obtain an estimated count as follows. Count representative portions of the plate receiving the highest dilution. Using a lighted colony counter with gridlines that are 1 cm apart (Fig. 1.3) assists in choosing a representative area of the plate to count. If there are fewer than 10 colonies/cm^2, then count 13 squares (7 consecutive horizontally and 6 consecutive vertically; see Fig. 1.5 for an example plate-counting field). The sum of the 13 squares multiplied by 5 equals the estimated count per 65-cm^2 plate (the size of a typical Petri dish). If there are more than 10 colonies/cm^2, count 4 representative squares and multiply the average by 65 to give the estimated count per plate. If there are greater than 100 colonies/cm^2, then record the count as >6500 CFU/plate. In all cases, Eq. 1.2 is used to obtain the CFU/g or CFU/ml (estimate). Never report the final count in the food sample as "too numerous to count" (TNTC). Plates with different surface areas, such as the Petrifilms, are counted using the same principle. Count a minimum of 4 squares and average those counts; then multiply that count by the area of the plate being used.

5. *Plates with Spreaders* Count a chain of colonies not too distinctly separated as a single colony. If colonies can be distinguished, then it is not considered a spreader for counting purposes. If chains of colonies appear to originate from separate sources, count each chain as one colony. If the spreader is greater than 25% of the plate, report the results as spreaders (Spr) rather than as a number.

REFERENCES

Difco Laboratories. 1998. *Difco Manual*, 11th ed. Difco Laboratories, Sparks, MD.

Harrigan, W. F. 1988. *Laboratory Methods in Food Microbiology*, 2nd ed. Academic, San Diego, CA.

Maturin, L. J., and J. T. Peeler. 1998. Aerobic Plate Count. In *U.S. FDA, Food and Drug Administration Bacteriological Analytical Manual*, 8th ed. (pp. 3.01–3.10). AOAC International, Gaithersburg, MD.

PRACTICING BASIC TECHNIQUE

OBJECTIVES

1. Practice aseptic technique.
2. Practice isolation and plating techniques: streaking, spread plating, and pour plating.
3. Apply the colony-counting rules.
4. Practice the use of a microscope.

PROCEDURE OVERVIEW

This exercise covers basic training in preparation of dilutions and plating techniques. Additionally, students will practice streaking for isolation of individual colonies and examining isolated colonies using the optical microscope. Aseptic handling of cultures and sterile media will be observed throughout this exercise. A flowchart of the work done in this laboratory exercise is shown in Fig. 1.6a. Students may work in groups of two, and prior experience in basic technique is shared. The exercise is completed in two laboratory sessions.

In preparation for this exercise, read the laboratory procedure carefully and annotate the items in Fig. 1.6a with appropriate experimental details. A copy of the annotated flow chart may be used as a guide while executing the laboratory exercise. Figure 1.6b is an example of how the flowchart may be annotated.

MEDIA

Plate Count Agar (PCA)

This is a general-purpose medium typically used for enumeration of organisms in food. Tryptone, glucose (also called dextrose), and yeast extract make this rich and complex medium suitable for growing a wide variety of microbes.

Peptone Water

This simple medium is used frequently as a diluent. It contains 1% peptone and 0.5% NaCl and provides a nearly isotonic environment for microorganisms. Some microbiologists prefer using peptone water over physiological saline (0.85% NaCl) as a diluent. Accidental contamination of the former is easier to detect because the solution turns turbid.

14 BASIC MICROBIOLOGICAL TECHNIQUES

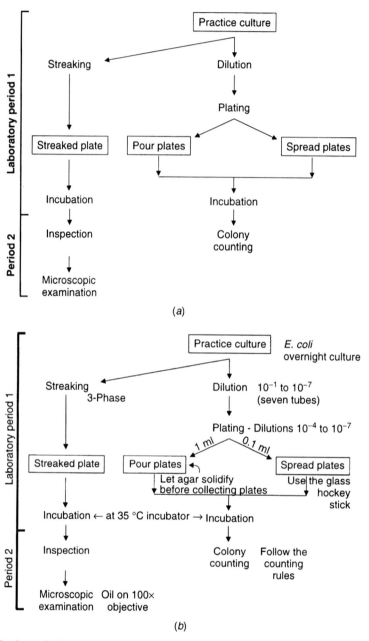

Fig. 1.6. Basic techniques practiced in the first laboratory exercise. Note that (*b*) is annotated.

Period 1 Dilution, Plating, and Aseptic Technique

Before doing this exercise, observe the instructor's demonstration of three-phase streaking, spread plating, pour plating, and aseptic transfer of media and cultures. To maintain aseptic technique, work in proximity to, but keep a safe distance from, the flame. Note also that the tubes, loops, and other items are flamed before and after each use. Tubes and plates should be kept closed except when actively in use.

MATERIALS AND EQUIPMENT

Per Student

- An overnight culture of *Escherichia coli*
- Eight 9.0-ml tubes of peptone water (dilution blanks)
- Five Petri plates containing PCA medium: one for streak plating and four for spread plates
- Four 15-ml tubes of molten PCA medium
- Four empty, sterile Petri dishes for pour plating
- Sterile pipettes, 1-ml
- Vortex mixer

Class Shared

- Waterbath maintained at ~50°C
- Incubator set at 35°C (±1°C)

PROCEDURE

An overview of the procedure applied in this exercise is shown in Figs. 1.6*a* and 1.6*b*. Additionally, the dilution and plating scheme is illustrated in Fig. 1.2.

Dilution

1. Prepare a rack of eight test tubes, each containing 9 ml of sterile dilution blank (peptone water). Label the tubes with the dilution factor (10^{-1}–10^{-8}) using an adhesive tape. Since the original tube is nondiluted, it may be given the dilution factor of 10^0.
2. Vortex the *E. coli* culture and transfer 1 ml into the first 9-ml test tube. This is a dilution factor of 10^{-1} (i.e., 1/10).
3. Vortex the 10^{-1} test tube and transfer 1 ml, using a new pipette, into the second 9-ml test tube. Vortex the second tube.
4. Repeat these steps to dilute the sample until 10^{-8} total dilution is reached (a total of eight test tubes).

16 BASIC MICROBIOLOGICAL TECHNIQUES

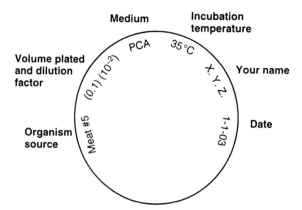

Fig. 1.7. Labeling the bottom of a Petri plate.

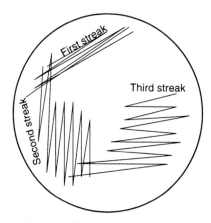

Fig. 1.8. Three-phase streaking.

Streaking

1. Label the bottom outer edge of a PCA plate with your name, the date, the medium, the organism, and the incubation temperature (Fig. 1.7).
2. Complete a three-phase streak (Fig. 1.8) of the nondiluted culture on the agar plate as follows:
   a. Vortex the culture. Flame the inoculating loop and allow it to air cool. Dip the loop into the culture (don't forget to flame the lip of the tube after uncapping and before recapping the tube).
   b. Beginning at the outer edge of the agar, move the loop in a zigzag pattern toward the center of the plate, taking care to not overlap lines. Flame the loop.
   c. Do the secondary streak by taking the flamed and cooled loop and crossing back into the primary streak 2–3 times to pick up organism and then zigzagging to spread the organism. Other than the first 2–3 times, do not go

back into the primary streak area. This portion of the streak uses half of the remaining portion of the plate.

    d. Repeat for the tertiary streak. This uses up the remaining portion of the plate.

3. Replace the plate lid, invert the plate, and incubate at 35°C for 24 hr.

Spread Plating

1. Label the bottom outer edge of four PCA plates with your name, the date, the medium, the organism, and the incubation temperature. Include the dilution factor and volume plated (Fig. 1.7).
2. Spread the dilute culture on four PCA plates (Fig. 1.2) as follows:
   a. Mix (using the vortex mixer) the 10^{-4} dilution tube of *E. coli* culture and dispense 0.1 ml onto the surface of the agar.
   b. Dip the bent end of the glass spreader into the alcohol jar, remove the spreader, and pass it quickly through the flame to allow the remaining alcohol to catch fire. Notice that alcohol sterilizes the spreader and flaming does not heat the spreader enough to sterilize it; it is done only to remove excess alcohol. Open the plate cover partially and spread the culture onto the agar surface using the sterilized spreader. Rotate the plate during the spreading to ensure an even distribution of the culture. Spread evenly by making sure the spreader reaches the edges as well as the center of the plate. Alcohol dip the used spreader and them flame the excess alcohol.
   c. Repeat using the 10^{-5}, 10^{-6}, and 10^{-7} dilutions.
3. Invert the plates and incubate at 35°C for 24 hr.

Pour Plating

1. Label the bottom outer edge of four sterile empty plates with your name, the date, the medium, the organism, and the incubation temperature. Also include the dilution factor and volume plated (Fig. 1.7).
2. Prepare four pour plates (Fig. 1.2) as follows:
   a. Mix (using the vortex mixer) the 10^{-5} dilution tube of *E. coli* culture and dispense 1 ml into the bottom half of the appropriately labeled Petri plate.
   b. Pour the molten agar from the PCA tubes provided into the plate. Pour enough agar to cover approximately two-thirds of the bottom of the plate. Gently slide the plates on the bench in a figure-8 motion, about three times, to mix the culture into the medium and spread the agar.
   c. Repeat using the 10^{-6}, 10^{-7}, and 10^{-8} dilutions.
   d. Allow agar to solidify before moving or inverting the plates.
3. Invert the plates and incubate at 35°C for 24 hr.

Period 2 Counting Microbial Populations in Culture

During the second period of this exercise, incubated plates are inspected, colonies developed during incubation are counted, and counting rules are applied. Selected colonies will be examined microscopically and results are recorded in the provided tables.

MATERIALS AND EQUIPMENT

Per Student

- Incubated plates (from last period)
- Plate that has been streaked with *Bacillus subtilis* culture and incubated at 35°C for 48 hr (provided by laboratory instructor)
- Colony counter
- Optical microscope with oil immersion lens
- Microscope slides and coverslips
- Microscope oil and material to clean oiled lens

RESULTS

Streak Plate of *E. coli*

1. Examine the streaked plate. Look for well-isolated colonies. Notice the streaking phase that produced most of the isolates. Look for any contaminants on the plate.
2. Prepare a wet mount and examine microscopically, as explained later.
3. Determine if there is a need to improve the procedure.
4. Record observations.
5. Dispose of the plate in the biohazard container.

Spread Plates

1. Judge the evenness of spreading by observing the distribution of colonies on the plates. Compare the appearance of the plates of different dilutions.
2. Notice the relatively uniform colony morphology compared to that present in pour plates.
3. Count CFU/ml of original culture by applying the colony-counting rules. Ability to visualize colonies is enhanced with the aid of a lighted colony counter (e.g., Quebec colony counter, Fig. 1.3). To minimize errors, mark the colonies on the back of the plate as they are counted using a marker.
4. Record results in Table 1.2.
5. Dispose of plates in the biohazard container.

TABLE 1.2. Counts of *Escherichia coli* Colonies on Plate Count Agar Petri Plates and Calculation of Count of Bacterium Population in Sample Tube

Method	Number of Colonies					Count (CFU[a]/ml)
	10^{-4}	10^{-5}	10^{-6}	10^{-7}	10^{-8}	
Spread plating					X[b]	
Pour plating	X					

[a] Colony-forming units.
[b] Dilution not plated.

Pour Plates

1. Judge the evenness of mixing by observing the distribution of colonies in and on the agar. Compare the appearance of the plates of different dilution.
2. Notice that colonies of the same microorganism can have different morphologies depending on their location in the agar. Colonies at the air–agar and plate–agar surfaces are different that those imbedded in the agar layer.
3. Count colonies at each dilution and record the results in Table 1.2. Mark the colonies, on the bottom of the plate, as they are counted using a marker.
4. Calculate the CFU/ml of original culture by applying the colony-counting rules. Record results in Table 1.2 and write down a sample of your calculations.
5. Dispose of plates in the biohazard container.

(***Attention:*** For this exercise, count all plates and mark, using footnotes, the dilutions that were used in CFU/ml calculations. In subsequent exercises, only colonies on plates suitable for calculating the CFU/ml or CFU/g are counted.)

Microscopic Examination

Wet Mount

1. Prepare a wet mount of a colony from the streak plates (*E. coli* and *B. subtilis*):
   a. Place a small droplet (one loopful) of water on the slide.
   b. Using a sterile loop, touch the colony to pick up microorganisms. Do not use the entire colony.
   c. Mix the loop's load into the water droplet.
   d. Place a coverslip over the suspensions.
2. Observation of slides:
   a. Examine the wet mounts using a bright-field optical microscope. Note that oil must be used with the 100× lens. Do not get oil on any other lens.
   b. Compare microorganisms in the two cultures (e.g., shape and size).
   c. Record your comments.

Gram Staining

1. Transfer a portion of well-isolated colony from the spread plate of *E. coli* and mix with a small drop of water (using the inoculation loop) onto a microscope slide.
2. Repeat the first step using a colony from the *B. subtilis* streak plate.
3. Emulsify and spread out the bacterial mass in the water drop using the loop.
4. Let the smears air dry.
5. Heat fix the smears by passing the slide quickly over a burner flame.
6. Gram stain the smears as follows:
   a. Cover the smears with crystal violet (primary stain). Rinse the stained smear gently with water after 60 sec of staining using a squeeze bottle. Holding the slide with a clothespin will help keep hands stain free.
   b. Cover the smears with Gram's iodine and rinse with water after 60 sec of treatment.
   c. Rinse with the decolorizer (acetone–alcohol mixture) for 2–3 sec, then rinse with water.
   d. Cover the smears with safranin (secondary stain) for 60 sec, then rinse with water.
   e. Carefully blot the slide dry using bibulous paper.
6. Examine microscopically the smears on the slide. Record observations regarding the shape and arrangement of the cells.
7. Record the interpretation of the Gram stain result. Gram-positive cells will appear blue to purple and gram-negative cells will appear pink to red.

PROBLEMS

1. What is the main objective of using a streak plate?

2. What are the advantages and disadvantages of pour plating and spread plating?

3. A microbiological culture was decimally diluted, 1-ml aliquots of selected dilutions were plated on duplicate plates of PCA medium, plates were incubated at 35°C for 48 hr, colonies on the plates were counted, and results were recorded as follows:

	Colonies	
Dilution	Plate 1	Plate 2
10^{-1}		Not counted
10^{-2}	489	520
10^{-3}	47	54
10^{-4}	2	8
10^{-5}	1	0

   (a) Calculate the CFU/ml in the original culture.
   (b) Why is the count reported as CFU/ml rather than cells/ml?

4. A sample of raw milk was decimally diluted, 0.1-ml aliquots of selected dilutions were plated on duplicate plates of PCA medium, plates were incubated at 25°C for 48 hr, colonies on the plates were counted, and results were recorded as follows:

	Colonies	
Dilution	Plate 1	Plate 2
10^{-1}		Not counted
10^{-2}	630	591
10^{-3}		Not counted
10^{-4}	5	3

   (a) Calculate the CFU/ml of milk.
   (b) How should this procedure have been modified to produce a more accurate count?

5. You are given a plate that has a large number of colonies (see the accompanying picture). Calculate the estimated original microbial count in the food sample, knowing that the plate was prepared by pour plating at a dilution of 10^{-2}. Show your calculations.

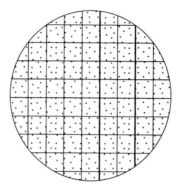

6. To determine yeast and mold counts in pasteurized apple cider, a sample (11 ml) was mixed with 99 ml diluent and aliquots (1 ml each) of the mixture were plated on four Petri plates containing antibiotic PCA medium. The plates were incubated at 25°C for 72 hr. When the plates were inspected in preparation for colony counting, no colonies were found on any of the four plates. Determine the yeast and mold CFU/ml of cider.

7. Why do gram-positive and gram-negative bacteria produce different colors with Gram staining. Explain the chemistry of this staining process.

PART II

FOOD MICROBIOTA

It is important for the food microbiologist to recognize that each food has its own characteristics and these characteristics determine what microbes are found in that food. Dry foods do not commonly contain significant numbers of organisms with limited ability to survive dry conditions, and sugary foods are not likely to be populated with microbes with low tolerance to osmotic pressure. Foods requiring elaborate preparation are more likely to have greater microbial diversity than foods that receive very little handling. Therefore, food properties and processing history determine, to a great extent, the types of microorganisms prevalent in that food. In other words, foods are likely to dictate their own microbiota.

Microbiota in raw food originates mainly from the production environment (i.e., the farm). Handling of this raw food during harvesting and shipping adds to the diversity of microbial contaminants. Food processing is usually designed to decrease or eliminate the microbial population in the food. While this is true for pasteurization and commercial sterilization of food, processing through fermentation is an exception to this rule. Fermentation entails adding beneficial microorganisms, in the form of a starter culture, to a food or food ingredients and incubating the mixture at optimum conditions for growth and multiplication of this culture. These intentionally added microorganisms are commonly well represented in the microbiota of the finished product. Postprocessing handling may increase the diversity of the microbial population of food through introduction of microorganisms commonly carried by humans. Therefore, the food we eat ranges from virtually sterile products (e.g., canned food) to products containing millions of living microbial cells (e.g., Cheddar cheese or yogurt). Figure II.1 shows some of the microorganisms that may be encountered in food. These are grouped based on characteristics readily observable by conventional microbiological techniques. The listing in Fig. II.1 includes microorganisms that (a) are beneficial, such as lactic acid bacteria and yeasts; (b) cause spoilage, such as *Pseudomonas* spp. and molds; and (c) are pathogenic, such as *Salmonella* spp. and *Listeria monocytogenes*.

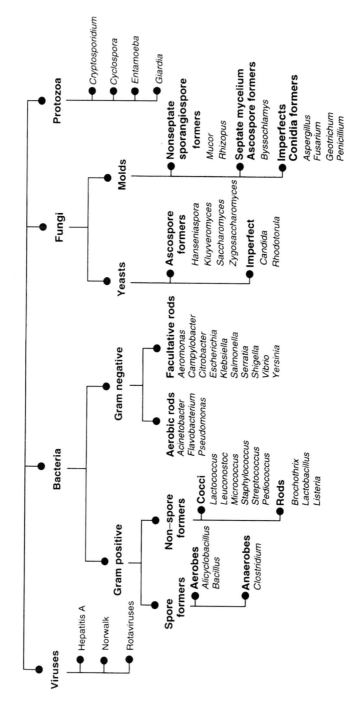

Fig. II.1. Selected foodborne microorganisms and viruses of significance in food. Note that only genera are presented. (See Chapter 3 for a detailed classification of foodborne fungi.)

This manual includes laboratory exercises designed to enumerate food microbiota (i.e., total plate count) or its subgroups (e.g., yeasts and molds) and to detect specific microorganisms, namely pathogens. In addition to enumerating or detecting microorganisms of concern in food, each exercise gives the student an opportunity to practice a new analytical technique. Each chapter title includes, in parentheses, the newly introduced technique. Description of relevant microbiological media are included in each chapter and detailed compositions of these media are listed in Appendix C.

Food microbiota are commonly grouped into categories with some uniform properties. Each category is enumerated based on the ability of member species to survive selective conditions or grow on a selective medium. In this part of the manual, total yeasts and molds, coliforms, and mesophilic spore formers are counted. Food processing or storage environment also will be sampled and the environmental microbiota will be assessed. When examining food microbiota, it is necessary to be familiar with the product sampled. Combining an analyst's observation about the characteristics of the microbiota of food with its properties and processing history may help assess the product's quality and safety.

CHAPTER 2

TOTAL PLATE COUNT
COMPARING HOMOGENIZATION METHODS AND PLATING MEDIA; APPLYING COLONY-COUNTING RULES

INTRODUCTION

Quality of food is related to its physical, chemical, microbiological, and sensory properties. The microbiological quality can be roughly evaluated by analyzing the food for total plate count (TPC). In some foods, high TPC may indicate poor quality. Food which appears normal may have a high TPC, indicating that the food is about to spoil. Testing food for TPC involves homogenizing the sample, making serial dilutions of the homogenized sample, plating on plate count agar, incubating plates aerobically at 35–37°C for 48 hr, enumerating colonies on plates, and calculating the CFU/ml or CFU/g of food (Fig. 2.1). Although TPC is a quick and efficient method to obtain a generalized idea about the microbiological quality of food, this test has these limitations:

- Fermented foods (e.g., Cheddar cheese) naturally contain a high microbial load, and thus TPC cannot be used to evaluate their general microbiological quality.
- The plating medium (plate count agar) does not ideally support the growth of fastidious microorganisms; thus these microorganisms are not well represented in the TPC.
- Incubation conditions favor growth of mesophilic aerobic bacteria; thus other categories, such as strict anaerobes, are overlooked.
- Molds and yeasts grow better at lower temperatures than those used in this test. Molds and yeasts also require a longer incubation period. Therefore, fungi are underestimated in the TPC.

This laboratory exercise is designed to train the students to analyze solid foods for total count and compare different methods of homogenizing and plating food

---

Food Microbiology By Ahmed E. Yousef and Carolyn Carlstrom
ISBN 0-471-39105-0 Copyright © 2003 by John Wiley & Sons, Inc.

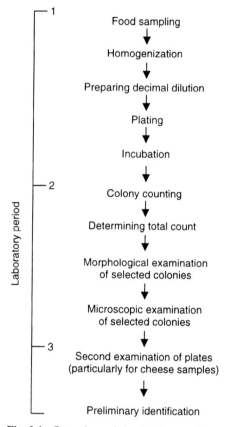

Fig. 2.1. Overview of the TPC procedure.

samples. Additionally, foods with inherently different microbial profiles and population sizes will be analyzed.

OBJECTIVES

- Determine and compare the TPC of different foods.
- Evaluate the factors that may influence TPC determination:
  (a) Methods of food sample homogenization (blending vs. stomaching)
  (b) Plating media (agar medium in Petri plates vs. Petrifilm)
- Practice applying colony-counting rules.
- Become familiar with the morphology of microorganisms commonly isolated from food.

PROCEDURE OVERVIEW

Foods that differ appreciably in properties and microbiota are suitable for analysis in this laboratory exercise. Therefore, cheese and raw meat are suggested for TPC

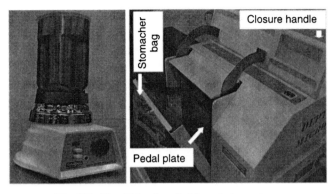

Fig. 2.2. Food blender (left) and stomacher.

determination. Samples of both food types will be prepared for analysis using either a blender or a stomacher (Fig. 2.2). The resulting homogenized samples will be diluted appropriately, depending on the microbial load expected in the food. Selected decimal serial dilutions are plated on Petrifilms or spread plated on PCA. Plates receiving meat samples will be incubated at 35°C for 48 hr, while those containing the cheese sample will be incubated at 30°C for 3–4 days. Colonies appearing on incubated plates are enumerated and CFU/ml or CFU/g of food is calculated. Selected colonies representing the microbiota of both foods will be examined and colony and cell morphologies will be determined.

This exercise will be completed in three laboratory periods, as shown in Fig. 2.1. During the first period, students will homogenize the food samples, prepare dilutions, plate on different media, and incubate the inoculated plates. Incubated plates will be inspected during the second period and counts are determined. Colony and cell morphology will be inspected during the second and the third periods.

Students will work in groups of two. Each group will sample one of two foods provided. Within each group, one student will homogenize the sample with the stomacher and the other with the blender. Each student will complete the analysis of the homogenized sample and share results with the partner. Results of all groups will be pooled in a table and shared with all students in the class.

In preparation for the laboratory, read the laboratory procedure carefully and annotate the items in Fig. 2.1 with appropriate experimental details (see Fig. 1.6*b* as an example). The annotated flowchart may be used as a guide while executing the laboratory exercise.

Food Handling

The following guidelines should be observed in all exercises in the manual:

- Foods should be kept in the laboratory refrigerator until just prior to laboratory start.
- The laboratory instructor should transfer the contents of the food's retail package aseptically into a suitable sterile plastic container (with cover) and mix thoroughly with a sterile spatula. Some foods require grinding or shredding

before samples can be taken. Such sample preparations should be done under aseptic conditions.
- The food package should be inspected visually for integrity or abnormalities.
- The package label should be read. Label information (e.g., package weight or volume, sell-by date, and food constituents) should be recorded and used in results interpretation.
- In most dilution schemes, 11 g (±0.1 g) of food sample is weighed. The sample should be weighed into a sterile stomacher bag (for stomacher samples) or into a sterile Petri dish (for blender samples). The bottom of the stomacher bag may be spread into a cup shape to support the bag on the scale. This facilitates weighing the sample directly into the bag. Note that it is crucial that sample contamination is minimized or eliminated by proper and prompt handling and weighing. Students who spend extra time weighing 11 g accurately (e.g., 11.00 g) are more likely to get their samples contaminated.

Food Homogenization

Procedures for enumeration of microorganisms vary greatly with the type of food analyzed. Sampling and homogenization before analysis, for example, are simpler in liquid than in solid foods. While liquid food samples are manually mixed before analysis, solid foods commonly require mechanical stirring (homogenization) in a suitable diluent to break food clumps and release microorganisms from the food matrix. Methods of homogenization may vary in ability to recover entrapped microorganisms. Blenders and stomachers (Fig. 2.2) are commonly used to prepare the food sample for analysis. The revolving blades of the blender divide the food into small particles and mix them with the diluent. The same goal may be achieved using a stomacher. The stomacher is a mechanical device that agitates a food sample placed in a sterile plastic bag. The back-and-forth mechanical action of the stomacher paddles mimics the stomach action (hence it is called a stomacher) and helps incorporate the sample particles into the diluent.

Preparing Dilutions

In preparation for plating, homogenized food samples are appropriately diluted, usually decimally, and some dilutions are selected for plating on the microbiological media. The extent of dilution and selection of the dilutions to be plated depend greatly on the analyst's expectations of the size of the microbial population in the food. The larger the size of this population, the greater the degree of sample dilution required. In this laboratory exercise, the dilution scheme, shown in Fig. 2.3, is suggested for analyzing meat and cheese.

Counting

After plates are incubated, colonies are counted to determine the microbial population in the analyzed sample. The colony-counting rules discussed in Chapter 1 will be followed. Figure 2.4 shows how to apply the colony-counting rules using a decision tree. A preliminary estimate of the number of colonies on different plates facilitates applying the counting rules in an efficient manner.

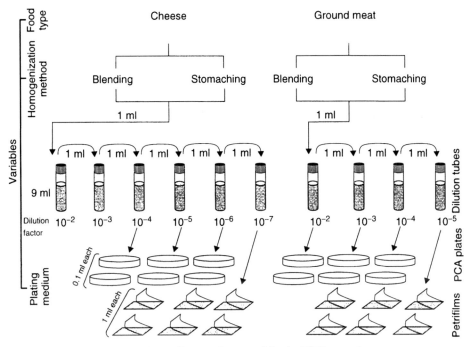

Fig. 2.3. Overall procedure used in the TPC exercise.

The population is commonly estimated as the number of colony-forming units per unit weight or volume of sample (i.e., CFU/g or CFU/ml). A colony could result from a single viable cell, a clump of cells, or a multicellular piece of mycelium. A single viable spore or a clump of spores also may form a single colony. The TPC, therefore, is an indication of the size of the population but it is not an accurate determination of the number of microbial cells in the sample. Samples are homogenized not only to liberate microbial cells from the food matrix but also to disperse clumps into smaller units or single cells.

Examination of Colonies

It is sometimes beneficial to gather information about the types of colonies resulting from plating the food sample. Note that in pour plating, a single species may produce more than one type of colony morphology, depending on the position of the colony in the agar medium. Surface colonies tend to be round while colonies embedded in the agar may show an oval or a star shape. Differences in morphology of colonies of a given microorganism are minimal in the case of spread plating or when pour-plated samples are overlayed with an additional quantity of the agar medium. In this laboratory exercise, spread plating will be used and, therefore, colony morphology will be relatively easy to interpret.

Microscopic examination provides additional information about the morphology of the cells that formed these colonies. If colonies of a given morphology are predominant, this stimulates the curiosity of the analyst and the food processor who may become interested in isolating and identifying a potentially prevailing microor-

32 TOTAL PLATE COUNT

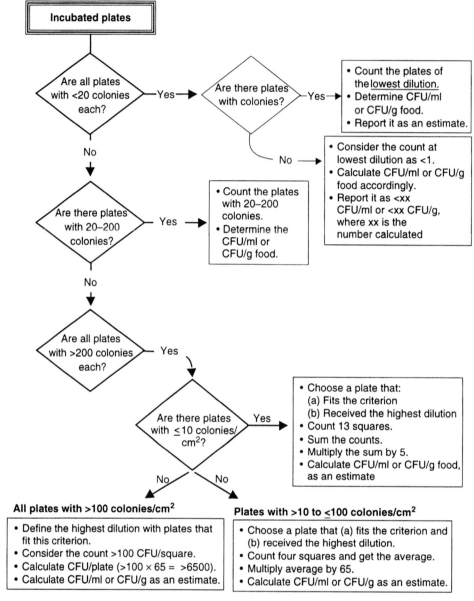

Fig. 2.4. Decision tree when applying colony-counting rules.

ganism in the food. In fact, examining the colony and cell morphology is an essential preliminary step in identifying isolates from food. For identification of a microorganism, the isolate should be additionally tested for biochemical, physiological, serological, or genetic traits (see later chapters on identification of pathogens in foods). It is, therefore, important to recognize that morphological testing, as done in this exercise, is not sufficient to identify the microorganism represented by the examined colonies.

MEDIA

Plating method and the medium used may affect the outcome of the analysis. The homogenized sample, with or without further dilutions, is mixed with the agar medium using a pour-plating or spread-plating technique. Plate count agar (PCA) medium (see Chapter 1 for medium description) will be used in this exercise.

A newer, non–Petri dish plating system will be practiced. It allows the analyst to mix the sample with a dehydrated medium on plastic film (Petrifilm) without the need to prepare and handle molten agar media. The Petrifilm plate consists of a layer of nutrient gel on a paper backing covered by a thin, flexible film. The gel in aerobic count Petrifilm plates also contains an indicator dye—tetrazolium. Living cells reduce the tetrazolium, causing the color to change to red. This reduction does not affect the viability of the cell. Therefore, colonies on aerobic count Petrifilms are readily identified as red spots in the plating area. All red dots, regardless of size, should be counted as colonies. By careful use of a needle, colonies from Petrifilms may be transferred to other media for further testing or isolation. The area outlined by the Petrifilm spreader is $20\,cm^2$, with each square on the film being $1\,cm^2$.

REFERENCES

Morton, D. R. 2001. Aerobic Plate Count. In F. P. Downes and K. Ito (Eds.), *Compendium of Methods for the Microbiological Examination of Foods*, 4th ed. (pp. 63–68). American Public Health Association, Washington, DC.

Smith, L. B., T. L. Fox, and F. F. Busta. 1985. Comparison of a Dry Medium Culture Plate (Petrifilm SM Plates) Method to the Aerobic Plate Count Method for Enumeration of Mesophilic Aerobic Colony-Forming Units in Fresh Ground Beef. *J. Food Prot.* 48: 1044–1045.

Yousef, A. E., E. T. Ryser, and E. H. Marth, 1988. Methods for Improved Recovery of *Listeria monocytogenes* from Cheese. *Appl. Environ. Microbiol.* 54:2643–2649.

Period 1 Sample Homogenization and Plating on Different Media

In this laboratory session, recovery of microorganisms from the food matrix will be done using two methods of homogenization, blending and stomaching. Additionally, food samples will be plated on an agar medium and Petrifilm plates (Fig. 2.3). Counts from different homogenization methods and plating systems will be compared in subsequent laboratory periods.

FOOD

Fresh ground beef
Cheese (Cheddar or blue)

MATERIALS AND EQUIPMENT

Per Pair of Students

- Stomacher bag
- Blender jar
- Two 99-ml bottles of peptone water
- Eight or twelve 9-ml tubes of peptone water, for analyzing meat and cheese, respectively
- Twelve PCA plates (agar medium prepoured 24–48 hr before use)
- Twelve aerobic count Petrifilms

Class Shared

- Scale for weighing food samples (e.g., a top-loading balance with 500-g capacity)
- Blender
- Stomacher
- Incubator, set at 30°C
- Incubator, set at 35°C

PROCEDURE

Sampling

A food retail package of ~500 g may be considered a sample. Samples should be obtained as described under the heading "Food Handling."

Homogenization

Blender Sample

1. Weigh 11 g of the food sample into a Petri dish. Transfer the sample into the blender jar.

2. Add 99 ml sterile peptone water to the blender jar. For homogenizing cheese samples, it is preferable to use a diluent containing citrate (2% trisodium citrate) at 40°C.
3. Homogenize the sample by blending for 2 min. The content of the blender (i.e., the homogenized sample) is the 1/10 (10^{-1}) dilution of the original sample.

Stomacher Sample
1. Weigh 11 g of the food sample into the stomacher bag.
2. Add 99 ml sterile peptone water to the stomacher bag. For homogenizing cheese samples, it is preferable to use a diluent containing citrate (2% trisodium citrate) at 40°C.
3. Homogenize the sample by stomaching for 2 min. The content of the stomacher bag (i.e., the homogenized sample) is the 1/10 (10^{-1}) dilution of the original sample.

Preparation of Dilutions

Prepare the following additional dilutions of the homogenized food sample using the 9-ml peptone blanks (Fig. 2.3):

Meat 10^{-2}, 10^{-3}, 10^{-4}, and 10^{-5}
Cheese 10^{-2}, 10^{-3}, 10^{-4}, 10^{-5}, 10^{-6}, and 10^{-7}

(*Note:* These dilutions are chosen based on expected counts in retail meat and cheese in the United States.)

One student of the group does a complete set of dilutions for the stomached sample and the other does a complete set for the blended sample.

Plating

Plate Count Agar Use the following dilutions for plating: 10^{-2}, 10^{-3}, and 10^{-4} for meat and 10^{-4}, 10^{-5}, and 10^{-6} for cheese (Fig. 2.3).

1. Label duplicate Petri agar plates at each dilution (a complete set is six plates). Prepare a complete set for the stomached sample and another for the blended sample.
2. Pipette 0.1 ml of the dilutions into the agar plate. Prepare two plates per dilution (i.e., duplicate plating). Repeat this with the other two dilutions.
3. Spread plate the inoculum using a sterile bent glass rod spreader (hockey stick) as described in Chapter 1.
4. Invert the plates and tape the group's plates together.
5. Incubate meat sample plates at 35°C for 48 hr. Incubate cheese sample plates at 30°C. Cheese sample plates may be inspected and preliminarily counted (without opening the plates) after 48 hr of incubation but typically require 3–4 days of incubation for better recovery of lactic acid bacteria and development of fungal mycelia.

Petrifilms Use the following dilutions: 10^{-3}, 10^{-4}, and 10^{-5} for meat and 10^{-5}, 10^{-6}, and 10^{-7} for cheese.

1. Label duplicate Petrifilms at each dilution (a complete set is six plates). Prepare a complete set for the stomached sample and another for the blended sample. This is a total of 12 Petrifilms. Petrifilms can be labeled by writing directly on the margin of the transparent side of the plate. Label them with the standard plate information.
2. Inoculate two Petrifilms at each of the three dilutions used as follows:
   a. Lift the cover film.
   b. Transfer 1 ml of the inoculum to the surface of the labeled Petrifilm. Keeping the pipette vertical should help prevent undesired spread of the inoculum.
   c. Carefully roll the film down to avoid entrapping air bubbles.
   d. Place the side of the spreader with the elevated ridge on top of the film over the inoculum. The indented circle will limit the area over which the inoculum spreads.
   e. Gently apply pressure on the spreader to uniformly distribute the inoculum over the entire circular area. Careful pipetting and gentle pressure should result in a uniform circle of inoculum.
   f. Lift the spreader and wait for 1 min for the gel to solidify.
3. Incubate the meat Petrifilm plates at 35°C for 48 hr. Incubate cheese sample plates at 30°C. These plates may be inspected after 48 hr of incubation but typically require 3–4 days of incubation for complete development of cheese microbiota.

Period 2 Counting Colonies and Examining Morphology

In this laboratory period, the counts of microbial population in the food sample (CFU/g) are determined following the colony-counting rules provided in the basic technique exercise (Chapter 1). Additionally, selected colonies will be examined and colony and cell morphologies are determined.

MATERIALS AND EQUIPMENT

Per Pair of Students

- Incubated plates
- Colony counter
- Gram stain kit
- Dissecting microscope
- Optical compound microscope

PROCEDURE

Colony Counting

1. Do a preliminary estimate of the number of colonies on all of the plates, regardless of whether the counts would be used in calculations.
2. Use the decision tree in Fig. 2.4 to determine the rules applicable to the plates prepared by the group.
3. Record the colony counts in Table 2.1 using only plates that fit the criteria of the counting rules.
4. Determine the CFU/g of food.
5. Record the results as CFU/g in Table 2.2 and in a table provided by the laboratory instructor for class data.

Colony Morphology

1. Select and mark six morphologically different, well-isolated colonies from the plates. Number each of the colonies, including the number on the plate. Colonies do not all have to be from the same plate.
2. Colonies may differ by color, size, consistency, margin, or other properties. Try to choose as much variety as possible.
3. Examine the colonial morphology of the six isolates using a dissecting microscope (stereo-microscope). This will allow observing the colony in more detail.
4. Record the observations in Table 2.3.

Cell Morphology

1. Transfer a portion of each of the six marked colonies to microscope slides (two colonies per slide) as follows:

TABLE 2.1. Counts of Colonies on PCA[a] Plates and Petrifilms of Meat (or Cheese) Samples Homogenized by Blender or Stomacher

Dilution Factor	Sample Homogenized by Stomacher (colonies/plate)				Sample Homogenized by Blender (colonies/plate)			
	PCA		Petrifilm		PCA		Petrifilm	
	Plate 1	Plate 2	Plate 1	Plate 2	Plate 1	Plate 2	Plate 1	Plate 2

[a] Plate count agar.

TABLE 2.2. *(Add a complete title that describes the data, the food analyzed, and the factors studied. Adjust the table to correctly report the data collected)*

Total Plate Count (CFU[a]/g), Stomached Sample		Total Plate Count (CFU/g), Blended Sample	
PCA[b]	Petrifilm	PCA	Petrifilm

[a] Colony-forming units.
[b] Plate count agar.
(Include additional footnotes, if necessary.)

Sample calculation: Show how one of the four numbers in the table was calculated; include descriptive words but not a narrative of the process.

TABLE 2.3. *(Add a suitable descriptive title, including food analyed, what was observed, and what was determined)*

Colony Number	Gram Reaction[a]	Colony Morphology	Cell Morphology	Relevant Microorganism[b]

[a] Gram reaction is not relevant to fungi.
[b] Foodborne microorganisms with similar morphology. Data gathered in this laboratory are not sufficient to identify any of the isolates. These data, however, may be sufficient to make guesses about potential microbial groups in the food.
(Include additional footnotes, if necessary.)

a. Label each end of the slide with a colony number using a wax pencil.
  b. Put a small drop of water (using the inoculation loop) on the slide.
  c. Touch lightly the colony center with the tip of a sterile inoculation loop and transfer to the water drop on the slide. Do not use the whole colony.
  d. Emulsify and spread out the bacterial mass in the water drop using the loop.
  e. Let the smears air dry.
  f. Heat fix the smears by passing the slide quickly over a burner flame.
2. Gram stain each smear as follows:
  a. Cover the smear with crystal violet (primary stain). Rinse with water after 60 sec of staining. Do not squirt water directly on the smear; rather, let the water run down over the smear. Holding the slide with a clothespin will help keep hands stain free.
  b. Cover the smear with Gram's iodine. After 60 sec, rinse with water.
  c. Rinse with acetone–alcohol (decolorizer) for 2–3 sec, then rinse with water.
  d. Cover the smear with safranin (secondary stain) for 60 sec, then rinse with water.
  e. Carefully blot the slide dry using bibulous paper.
3. Examine microscopically the smears on the slides. Record observations, regarding the shape and arrangement of the cells, in Table 2.3.
4. Record the interpretation of the Gram stain results. Gram-positive cells will appear blue to purple and gram-negative cells will appear pink to red. It is also possible that results could appear gram variable. Keep in mind that while fungi will absorb the stains used in the Gram staining process, results are not reported either gram positive or gram negative.

PROBLEMS

1. Food manufacturers may use TPC for different reasons.

   (a) Why would the manufacturer of the food, analyzed in this laboratory, conduct the TPC testing?

   (b) Provide at least three reasons why the TPC test is only an estimate of total microbial load in the particular food you analyzed.

2. For the food that your group tested in this laboratory, list the following information:

   (a) Method of packaging
   (b) Storage conditions

   Describe how these two factors may influence the type of microorganisms in the food you analyzed.

3. How do analysts determine which food sample dilution range to prepare and which of these dilutions to plate on agar media? Give an example and include the calculations involved in this decision.

4. In this laboratory exercise, plates receiving meat samples were incubated at 35°C for 48 hr, while those containing the cheese sample were incubated at 30°C for 3 days.

   (a) What is the logic for using different incubation conditions for these two foods?
   (b) What microbial groups are not well represented in each of these two counts?
   (c) How significant are these poorly represented groups for the microbiological quality of the analyzed food?

5. Fill out Tables 2.1, 2.2, and 2.3. Note that Tables 2.2 and 2.3 need complete and descriptive titles. For example, "Counts for Blue Cheese" is not an acceptable title. A descriptive title might include the fact that the numbers in the table are for total plate count, that two homogenization methods and media types were used, and that a specific type of food is analyzed. The tables provided already have some footnotes in place as examples. Examine each table and determine if additional footnotes are required; look for things like abbreviations or places you wish to add a bit more information. A sample calculation is required for Table 2.2. A proper and complete sample calculation uses a single representative set of data for the calculation. It includes a word description of what the numbers are. It includes all necessary numbers and units. See Appendix A for a general table structure.

6. Microbial counts, as determined in this exercise, are reported as CFU/g (or CFU/ml) rather than cells/g (or cells/ml) of food. Why?

7. Compare the counts from the blender and stomacher that your group obtained. Describe how these different methods could yield different results.

TABLE 2.4. Average Microbial Counts in Ground Meat and Blue Cheese Resulting from Blender and Stomacher Homogenization Treatments When Plated on PCA[a] and Petrifilm Media

	Average CFU[b]/g			
	Blender		Stomacher	
Food	PCA	Petrifilm	PCA	Petrifilm
Beef	2.9×10^6	2.4×10^6	1.5×10^6	1.6×10^6
Cheese	1.4×10^8	2.4×10^7	1.1×10^8	2.1×10^7

[a] Plate count agar.
[b] Colony-forming units.

8. Compare the two methods of homogenization tested with regard to general ease of use and suitability for use in a quality control laboratory.

9. Compare the two media tested with regard to general ease of use and suitability for use in a quality control laboratory.

10. Compare the counts that your group obtained with the class average for the same type of food.

11. Using the data gathered from the whole class (class data), compare counts and contrast between the foods that the class tested. Explain differences in the counts.

12. For all the colonies that you examined, explain how you guessed the potential colony identity. If you could not determine a preliminary identity, what is the additional information required?

13. Suppose the class obtained the results in Table 2.4 when this exercise was executed.

    (a) What trends are apparent in the data?
    (b) What possible explanations are there for these trends?

CHAPTER 3

YEASTS AND MOLDS
USING SELECTIVE MEDIA; EXAMINING FUNGI MORPHOLOGY

INTRODUCTION

Fungi (singular, *fungus*) are eucaryotic heterotrophic organisms that possess a poorly differentiated body called a thallus. The thallus of fungi varies in complexity from a unicellular or filamentous structure to a complex treelike form. It lacks specialized tissues typical of higher plants such as a stem, leaves, and conducting tissue. Fungi reproduce asexually through spore formation or by budding. Some fungi also produce sexual spores, not as a reproductive but rather as a protective measure. Production of sexual and asexual spores is an important criterion in the classification of fungi (Table 3.1 and Fig. 3.1).

Compared to bacteria, fungi are larger in size and produce complex structures and their cells contain nuclei. Fungi are aerobic microorganisms and generally grow slower than bacteria. Molds grow into filamentous structures called mycelia (singular, *mycelium*) that are easy to detect visually on media or food. Yeasts are commonly unicellular organisms, and their colonies cannot be distinguished readily from those of bacteria. Yeast cells, however, are much larger than those of bacteria. Fungi tend to grow in the mesophilic-to-psychrotrophic range. They are generally osmotic and acid tolerant.

Foodborne Fungi

Fungi, which include molds and yeasts, are classified into subkingdoms on the basis of septation of mycelia and into phyla based on their ability to produce sexual spores and type of sexual spores produced (Table 3.1 and Fig. 3.1). A phylum is further divided into classes, orders, families, genera, and species (Fig. 3.1). Tables

---

Food Microbiology By Ahmed E. Yousef and Carolyn Carlstrom
ISBN 0-471-39105-0 Copyright © 2003 by John Wiley & Sons, Inc.

TABLE 3.1. Characteristics of Fungi Phyla

Phylum	Septate Hyphae	Asexual Sporulation	Sexual Sporulation	Comments
Chytridiomycota	–	Motile zoospores	Oospore	Lower fungi
Zygomycota	–	Nonmotile sporangiospores	Zygospore	Lower fungi Not xerophilic Not resistant to heat and chemicals Rare mycotoxin production
Ascomycota	+	Conidiospores	Ascospore	Higher fungi Mostly xerophilic Often resistant to heat and chemicals
Basidiomycota	+	Rare	Basidiospore	Higher fungi Mushrooms, puffballs, rusts, jelly fungi No significance in food spoilage
Deuteromycota	+	Conidiospores	None	Higher fungi Some xerophilic Not resistant to heat Some resistant to chemicals Production of mycotoxin common

3.2 and 3.3 list fungi of concern in food, with comments on their morphology and properties.

Fungi are ubiquitous microorganisms that may contaminate foods, equipment, and processing and storage facilities. Because they are so widespread, fungi are the major cause of food spoilage. Consequently, contamination of food with certain fungi may cause considerable economic losses to the food industry. Some molds are considered pathogenic to humans since they may produce toxins in food. Mold toxins, commonly referred to as mycotoxins, vary greatly in structure, properties, and toxicity. Mycotoxins include aflatoxins, which are highly carcinogenic; T-2 toxin, which causes alimentary toxic aleukia; and ochratoxin, which leads to kidney toxicity.

Some fungi are beneficial since they are used in food fermentations. These include baker's yeast (*Saccharomyces cerevisiae*), which is used in bread making, and *Penicillium* spp., the molds used in ripening of Camembert and blue cheeses. Additionally, *Saccharomyces* spp. are used in beer and wine fermentation. Other beneficial molds are used to produce valuable food ingredients. *Mucor* spp., for example, are the source of a microbial rennet. Rennet is a food ingredient containing a proteolytic enzyme that curdles milk and thus is used in cheese manufacture.

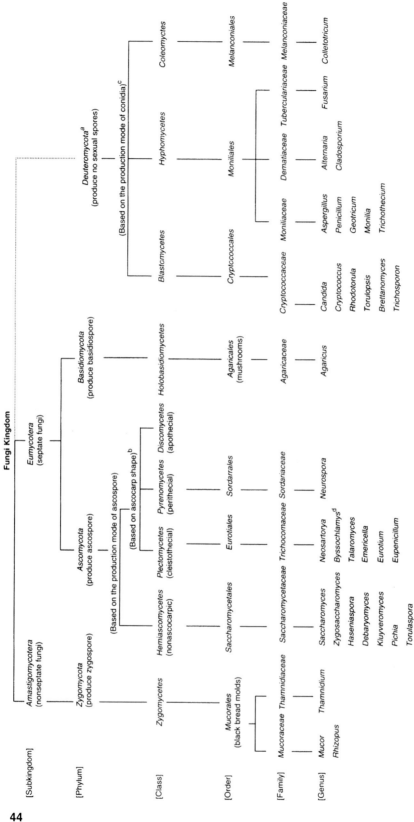

Fig. 3.1. Classification of fungal genera associated with food.

INTRODUCTION

TABLE 3.2. Characteristics of Food Spoilage Molds

Genus	Identification Characteristics	Microscopic Morphology	Important Properties Related to Food Spoilage
Zygomycota			
Mucor	Nonseptate hyphae Cottony colony Smooth, nonstriated sporangiospore Produce no rhizoids		Grow on refrigerated meat, cause defect called "whiskers" Black spot on frozen mutton Very common on bread
Rhizopus	Nonseptate hyphae Stolons, rhizoids Umbrella-shaped columellae Large sporangiospore with striated wall Dark sporangia containing dark to pale spores		Bread mold Watery soft rot of fruits Black spot on beef, bacon, frozen mutton
Thamnidium	Nonseptate hyphae Sporangia on highly branched structure Clusters of smaller sacs (sporangioles) present as well as sporangia (often on the same stipe)		Whiskers on beef
Ascomycota			
Neosartorya	White, cottony, fluffy colony White cleistothecia Colorless ascospores Gray, green conidia		Heat-resistant ascospores Spoil canned bottled fruit juice Not xerophilic
Byssochlamys	Cottony colony *B. fulva*: tannish yellow to olive green colonies *B. nivea*: white to cream colonies Absence of ascocarp, asci in open clusters		Heat-resistant ascospores Spoil canned bottled fruit juice
Eurotium	Bright yellow cleistothecia Pale yellow, oblate (may have ridges) ascospores Gray, green conidia		Xerophilic Spoilage in grape jam and jelly

TABLE 3.2. (*Continued*)

Genus	Identification Characteristics	Microscopic Morphology	Important Properties Related to Food Spoilage
	Deuteromycota		
Aspergillus	Black, brownish black, purple brown conidiophore Yellow to green conidia Dark sclerotia		*A. niger*: black rot on fruits and vegetables Yellow, green to black on large number of foods
Penicillium	*P. digitarum*: yellow green conidia *P. italicum* and *P. expansum*: blue-green conidia *P. camemberti*: grey conidia		Blue/green rots of citrus fruits Soft rots of apple, pear, peaches
Geotrichum	Arthroconidia formation White colonies Colorless conidia		Machinery mold Soft rot of citrus fruits, peaches Common in dairy products Some have strong odors
Monilia	Pink, gray, or tan conidia		Red bread mold Brown rot of stone fruits
Trichothecium	Long conidiophore, simple hyphae Ellipsoidal to pyriform conidia Conidia with a single lateral septum		Pink mold grows on fruits Common in grains, including barley, wheat, and maize
Alternaria	Large, brown conidia Club-shaped conidia		Brown to black rot in apple, figs, citrus fruits Found on red meats Field fungus: grows on barley and wheat
Cladosporium	Thick, velvety colony Green, olive green, dark-blue, black or brown colony Some lemon-shaped conidia Variously branched		Black spot on meat, beef Some spoil butter, margarine Field fungus: grows on barley and wheat
Fusarium	Cottony, pink, red, purple, brown colonies Sickle-shaped conidia Extensive mycelium Colorless conidia		Soft rot of figs Brown rot of citrus fruits, pineapples Field fungus: grows on barley and wheat Bacon, refrigerated meat spoilage, pickle softening

TABLE 3.3. Characteristics of Food Spoilage Yeasts

Genus	Identification Characteristics	Microscopic Morphology	Important Properties Related to Food Spoilage
	Ascomycota		
Saccharomyces	Multilateral budding Spherical spore Does not ferment lactose White or cream colonies Typical yeasty odor		*S. cerevisiae*: ubiquitous contaminant, sometimes fermentative spoilage of soft drinks, some strains preservative resistant
Zygosaccharomyces	Multilateral budding "Dumbbell"-shaped asci One to four ascospores per ascus Bean-shaped ascospores Strong fermenter of sugars		*Z. bailii*: highly resistant to preservatives, xerophile, capable of growth at $a_w = 0.8$,[a] at pH 1.8, heat-resistant ascospores, spoil tomato sauce, mayonnaise, salad dressing, soft drinks, fruit juices etc. *Z. rouxii*: grow at $a_w = 0.62$
Haseniaspora	Bipolar budding Lemon-shaped (apiculate) cell Two to four ascospores per ascus		Found on figs, tomatoes, citrus fruits, strawberries Grow in fruit juices
Debaryomyces	Multilateral budding Spherical or oval ascospore Pseudomycelium		Prevalent yeasts in dairy products Forms slime on wieners Grow in brine, cheeses *D. hansenii*: high salt tolerance, grow at $a_w = 0.65$, spoil yogurt, orange juice concentrate
Kluyveromyces	Multilateral budding Spherical spores Vigorous fermenter of sugars		Prevalent yeasts in dairy products *K. marxianus*: cheese spoilage
Pichia	Multilateral budding Four ascospores per ascus Pseudomycelium Arthrospores		Found on fresh fish, shrimp Grows in olive brine *P. membranaefaciens*: preservative resistant, film formation on olives, pickles, sauces

TABLE 3.3. (*Continued*)

Genus	Identification Characteristics	Microscopic Morphology	Important Properties Related to Food Spoilage
	Deuteromycota		
Candida	Cells are spheroidal, cylindrical, ovoid, or elongate Pseudomycelium formation		Common yeasts in fresh ground beef and poultry *C. krusei*: preservative resistant, form film on pickle, olives, sauces *C. parapsilosis*: spoil chese, margarine, dairy and fruit products
Cryptococcus	Multilateral budding Nonfermenter of sugars		Found on strawberries, other fruits, marine fish, shrimp, fresh ground beef
Rhodotorula	Multilateral budding Nonfermenter of sugars Orange or salmon pink color pigmentation Often mucoid colonies		Found on fresh poultry, shrimp, fish, beef Some grow on surface of butter *R. mucilaginosa*: spoil dairy products, occasional spoilage of fresh fruits
Brettanomyces	Multilateral budding Formation of ogival cells (pointed at one end, rounded at the other)		Spoil beer, soft drink, wine, pickles *B. intermidius*: prevalent species, grow at pH as low as 1.8 *B. bruxellensis*: off odors in beer, cider, soft drinks

[a] a_w = water activity.

The presence of a large population of yeasts and molds is undesirable in food, with the exception of mold-ripened cheese. A high count of yeasts and molds in food may indicate poor sanitation and handling, temperature abuse, inadequate processing, or postprocessing contamination. Food ingredients that were exposed to mold growth (e.g., flour made from moldy corn) may contain hazardous mycotoxins. Processing food made using such an ingredient may decrease the yeast and molds count substantially but not the level of the mycotoxin. In this case, analysis for mycotoxins is needed to assess the safety and quality of these foods. Fungi growing in acid food may consume acidic ingredients and thus raise the pH of the product. This may lead to conditions that support the growth of hazardous bacteria that were inhibited initially by the food's acidity.

OBJECTIVES

1. Determine and compare yeast and mold counts of different foods.
2. Compare different media used in counting yeasts and molds.
3. Become familiar with the macroscopic and microscopic appearance of foodborne yeasts and molds.

PROCEDURE OVERVIEW

Students will analyze fruit juices for yeasts and molds. Some groups will test a commercial juice and others will analyze a fruit juice prepared in the laboratory using a simple extraction technique. Yeasts and molds are aerobic microorganisms, and for their enumeration surface plating is preferred over pour plating. However, compared to pour plating, surface plating decreases the method's detection limit since only a 0.1-ml sample is spread on the plate. Methods to enumerate yeasts and molds in food include sample homogenization, plating on suitable selective media, incubation, and colony counting. Predominant colonies on the agar media will be examined microscopically and compared to known fungi. This exercise will be completed in three laboratory periods (Fig. 3.2). The first period of this exercise will be dedicated to extraction of juice from fresh fruits and preparation of plates of selective agar media. Media plates need to be prepared 24–48 hr before plating to allow drying of free water on agar surface. During the second period, students will prepare dilutions of juice samples, plate appropriate dilutions onto three selective media, and incubate the inoculated plates. During the third period, incubated plates will be inspected macroscopically (for colony morphology) and microscopically (for cell morphology). Students will compare their isolates with premounted samples of known yeasts and molds. Based on this comparison, the genus of some isolates may be hypothesized. Therefore, the variables studied in this exercise are food type (processed vs. fresh) and type of selective medium, that is, acidified potato dextrose agar (acidified PDA), antibiotic plate count agar (antibiotic PCA), and yeast and mold Petrifilm.

Each group will sample one assigned food. Members of the group will distribute the experimental work among themselves (e.g., one student prepares the dilutions and does the plating on acidified PDA and the other performs the plating on antibiotic PCA and Petrifilms). Data will be shared by both members of the group. Notice that some steps are common for all plating methods (e.g., preparation of dilutions) and are done once. Others will be repeated for each plating technique.

In preparation for the laboratory, annotate the items in Fig. 3.2 with appropriate experimental details (e.g, volume of tartaric acid added to 100 ml PDA) using a pencil. The annotated flowchart may be used as a guide while running the exercise in the laboratory. Deviations for the procedure should also be marked on the annotated chart.

MEDIA

Acidified Potato Dextrose Agar

Potato dextrose agar is a nonselective nondifferential medium. The rich potato infusion encourages growth of yeasts and molds. Addition of tartaric acid to PDA

50 YEASTS AND MOLDS

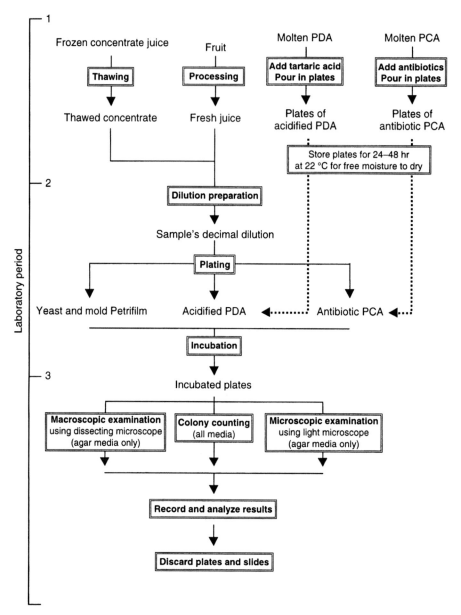

Fig. 3.2. Stages of testing for molds and yeasts in food.

lowers the pH of the medium to a level that inhibits growth of bacteria but permits growth of fungi. Unfortunately, some acid-tolerant bacteria may grow on this medium and thus cause overestimation of yeast and mold count.

Antibiotic Plate Count Agar

Plate count agar medium is described in Chapter 1. The addition of appropriate antibiotics makes this medium selective for fungi. Antibiotics such as tetracycline and chloramphenicol act against bacteria but will not inhibit the growth of eucaryotic yeasts and molds.

Petrifilms

The yeast and mold count Petrifilms are selective for yeasts and molds. They are similar to the aerobic count Petrifilms (Chapter 2) but also contain antibiotics (chlortetracycline and chloramphenicol) to create selectivity. Yeast and mold Petrifilms contain an indicator (bromo-chloro-indolylphosphate) that gives yeast colonies a blue-green color. These colonies will be small and defined, similar to those of bacteria on the aerobic count Petrifilms. Molds will be larger and more diffuse or fuzzy in appearance; their colonies (mycelium) are typically blue, from the dye in the medium, but may retain natural pigmentation (e.g., blue, black, yellow). Note that yeast and mold Petrifilms have a growth area of $30\,cm^2$.

REFERENCES

Deak, T., and L. R. Beuchat. 1996. *Handbook of Food Spoilage Yeasts*. CRC Press, Boca Raton, FL.

Guarro, J., J. Gene, and A. M. Stchigel. 1999. Developments in Fungal Taxonomy. *Clin. Microbiol. Rev.* 12(3):454–500.

Pitt, J. I., and A. D. Hocking. 1997. *Fungi and Food Spoilage*, 2nd ed. Blackie Academic and Professional, Cambridge, MA.

Szaniszlo, P. 2002. *Biology of Fungi*. Available: `http://www.esb.utexas.edu/mycology/bio341`.

Period 1 Preparing Food and Selective Agar Media

MATERIALS AND EQUIPMENT

Per Pair of Students

- Frozen concentrated citrus juice or a citrus fruit
- One bottle containing 100 ml molten PDA
- One bottle containing 100 ml molten PCA
- One tube containing 2.5 ml filter-sterilized tartaric acid (10%)
- One tube containing 2.5 ml antibiotic solution (antibiotic concentrations used are shown later)
- Twelve sterile, empty Petri dishes
- Clean container for receiving extracted juice

Class Shared

- Clean fruit juice extractor
- Waterbath, set at 50°C

FOOD PREPARATION

1. Transfer packages of the frozen concentrated citrus juice from the freezer to the refrigerator. Refrigerate the product 24–48 hr to allow thawing.
2. Cut fresh citrus fruit into halves. Extract juice using a manual juice extractor (Fig. 3.3). Receive extracted juice in a suitable clean container. Clean extractor surface before each use to minimize cross contamination between samples. Depending on the type of juice extractor used, a filtration process may be required. Refrigerate the juice in a closed, properly labeled container until analyzed in the subsequent laboratory period.

SELECTIVE AGAR MEDIA PREPARATION

Acidified PDA

Add tartaric acid to the molten PDA at ~48°C to reach a pH of 3.5. It has been determined that 1.85 ml of a 10% sterile solution of tartaric acid will decrease the pH of 100 ml PDA medium to 3.5. Mix gently to avoid introduction of air bubbles. Divide bottle contents among six Petri plates. Store plates at room temperature on bench counter for 24–48 hr to allow the free moisture on the agar surface to dry.

Antibiotic PCA

Measure 2 ml of the antibiotic solution and add into 100 ml of molten PCA medium (~48°C). Mix gently to avoid introduction of air bubbles. The antibiotic solution used

Fig. 3.3. Manual citrus juice extractor.

in this experiment was prepared as follows: 500 mg each of chlortetracycline HCl and chloramphenicol are dissolved in 100 ml of ethanol, filter sterilized, and the solution is stored in the refrigerator. Divide bottle contents among six Petri plates. Store plates at room temperature on bench counter for 24–48 hr to allow the free moisture on the agar surface to dry.

Period 2 Sampling, Dilution, and Plating

Juices should be mixed well before sampling, and the sample carefully agitated just before analysis. Dilutions of the food sample are prepared and plated on acidified PDA, antibiotic PCA, and yeast and mold Petrifilm. Plates are incubated right-side up (i.e., agar side down) at 22–25°C for 5 days.

MATERIALS

Per Pair of Students

- Thawed juice or laboratory-extracted juice
- Two 9-ml peptone water dilution blanks
- Six Petri plates containing the acidified PDA (prepared in first period)
- Six Petri plates containing the antibiotic PCA (prepared in first period)
- Four yeast and mold Petrifilms
- Alcohol jar
- Bent glass spreader
- Pipettes, 1 ml, and pipetter bulb

Class Shared

- Incubator, set at 25°C

PROCEDURE

Sampling and Preparing Dilutions

1. Inspect the food package and obtain label information. In case of the laboratory-extracted juice, record processing and storage conditions.
2. Shake the container to mix the contents, transfer a sample (~10 ml) aseptically into a sterile test tube for use by the group.
3. Prepare 10^{-1} and 10^{-2} dilutions of the sample using the 9-ml peptone blanks.

Plating on Acidified PDA

1. Dispense 0.1 ml of the 10^0 dilution (undiluted) onto two plates containing acidified PDA. Repeat for the 10^{-1} and 10^{-2} dilutions. This is a total of six plates.
2. Spread the inocula evenly on the plates using a sterile bent glass spreader.
3. Tape the group's plates together.
4. Incubate plates, right-side up, at 25°C for 5 days. The plates should remain undisturbed until counted.

Plating on Antibiotics PCA

1. Dispense 0.1 ml of the 10^0 dilution (undiluted) onto two plates containing the antibiotic PCA. Repeat for the 10^{-1} and 10^{-2} dilutions. This is a total of six plates.

2. Spread the inocula evenly on the plates using a sterile bent glass rod.
3. Tape group's plates together.
4. Incubate plates, right-side up, at 22–25°C for 5 days. The plates should remain undisturbed until counted.

Yeast and Mold Petrifilms

1. Transfer 1 ml of the diluted sample (10^{-1} and 10^{-2}) to the surface of the labeled Petrifilm. Keeping the pipette vertical should help prevent undesired spread of the inoculum. Do not inoculate Petrifilms with undiluted sample. Undiluted sample tends to produce artifacts on Petrifilms and thus will not be tested.
2. Carefully roll the film down to avoid entrapping air bubbles.
3. Place the indented side of the spreader on top of the film over the inoculum. Gently apply light pressure on the spreader to distribute the inoculum uniformly over a circular area. This limits spread of inoculum to a 30-cm^2 area.
4. Lift the spreader and wait for 1 min for gel to solidify.
5. Tape the group's films together.
6. Incubate at 22–25°C for 5 days.

Period 3 Counting Colonies and Examining Morphology

Precautions Exposure to yeasts and molds may cause health hazards. Some fungi cause infections, allergies, and intoxications. Therefore, care should be exercised when handling plates containing yeast and mold colonies or when taking samples for microscopic examination. It is advisable to wear gloves while sampling mold colonies by the adhesive tape.

MATERIALS AND EQUIPMENT

Per Pair of Students

- Incubated plates
- Lactophenol cotton blue stain
- Transparent adhesive tape
- Crystal violet
- Microscope

PROCEDURE

Colony Counting

Count the number of yeast and mold colonies. Yeast colonies will be those similar to bacterial colonies in appearance. Mold colonies will typically be larger and fuzzier. Overgrowth of mold may cause difficulties in counting. In this case, counting the colonies from the underside of the plate may be helpful. Use reasonable judgment in counting; if it looks like two different overlapping colonies, it probably is.

Record data in Table 3.4. After inspection of the plates and some preliminary counting and estimation of counts, only plates best suited for use in calculations are counted. Follow the counting rules as explained in Chapter 1. For example, if all of the most dilute plates have less than 20 colonies, all of the middle dilution plates have 20–200 colonies, and the most concentrated plates have over 200 colonies, only the middle dilution plates are counted. Include footnotes pertinent to this information in the table.

Calculate the CFU of yeasts and molds per milliliter of juice. Be sure to do calculations using only the appropriate countable plates.

Microscopic Examination of Colonies

Yeast

1. Pick three yeast colonies, representing different colony morphologies, from any of the plates. Transfer a portion of the colony to a drop of water on a microscope slide.
2. Emulsify and spread the mixture and heat fix the resulting smear.
3. Stain using a simple stain (i.e., crystal violet) for 60 sec.
4. Examine the cells using the oil immersion lens (100×). Draw the yeast's cellular morphology in Table 3.5.

TABLE 3.4. *(Add a descriptive title for these data, including food, media, and data obtained)*

Growth Medium	Replicate	Volume Plated	Number of Colonies			Count (CFU/ml)
			10^0	10^{-1}	10^{-2}	
Acidified PDA	1					
	2					
Antibiotics–PCA	1					
	2					
Yeast and mold Petrifilm	1		x[a]			
	2		x			

[a] Dilutions not plated on this medium.
(Add at least two footnotes clarifying the presentation of the data and abbreviations).

Mold

1. Place a drop of lactophenol cotton blue dye on a slide.
2. Gently touch the surface of a mold colony with the adhesive surface of a small piece of tape. Place the tape on the drop of dye. Observe the necessary precautions, to avoid health hazard, when executing this step.
3. Examine the slide using the microscope. Draw the mold's cellular morphology. Mycelia and reproductive structures will both be visible in a successful stain. It may be necessary to prepare several slides to determine how hard to press to transfer the mold to the adhesive tape properly.
4. Repeat these steps so that three mold colonies, preferably with different morphologies, are examined.
5. Record your observations in Table 3.5.

Demonstrations

Five side demonstrations (agar plates under dissecting microscopes and slides under light microscopes) are prepared by laboratory instructors. The organisms demonstrated are as follows:

- *Rhizopus* sp.
- *Penicillium* sp.
- *Aspergillus* sp.
- Budding yeast (*Saccharomyces cerevisiae*)
- A bacterium (e.g., *Escherichia coli*)

TABLE 3.5. *(Add a descriptive title for these data)*

Sample	Colony Morphology	Cellular Morphology[a]
Yeast (from juice sample: _____)	1.	
	2.	
	3.	
Mold (from juice sample: _____)	1.	
	2.	
	3.	
Budding yeast (*Saccharomyces cerevisiae*) [Demonstrated]		
Rhizopus sp. [Demonstrated]		
Penicillium sp. [Demonstrated]		
Aspergillus sp. [Demonstrated]		
Escherichia coli [Demonstrated]		

[a] Yeasts and bacteria observed at a magnification of 1000×; molds observed at a magnification of 100×.

Demonstrated Yeast

1. Examine the yeast and the bacterium colony morphology on the agar media plates, and record your observations in Table 3.5.
2. With the dissecting microscope examine the yeast and the bacterium colony morphology on the agar media plates after removing the plate lid. Look at colony characteristics such as color, margin, shape, and size. Record your observations.
3. Use the compound light microscopes to examine the specimens of the yeast and the bacterium. Draw each of the demonstrations viewed.
4. Compare the microscopic appearance of the demonstrated yeast with that of the yeasts from the food sample that was prepared in the class.

Demonstrated Mold

1. Examine the mold colony morphology and record your observations in Table 3.5.
2. Examine the mold mycelium under the dissecting microscope as follows:
    a. Place the mold plate, with agar side down, on the stage of a dissecting microscope.

    b. Leave the lid in place to prevent spore dispersion into the room.
    c. Examine the mycelium and record your observation.
    d. Remove the plate from the stage when done to reduce water condensation buildup on the lid.
 3. Use the compound light microscope to examine the molds and describe the mycelium structure (e.g., septate or nonseptate and types and arrangements of spores). Draw each of the demonstration views.
 4. Compare the appearance of the demonstration views with the appearance of the mold from the food sample that was prepared in the class.

60 YEASTS AND MOLDS

PROBLEMS

1. What is the practical value of the data you gathered in this laboratory exercise for a fruit juice manufacturer?

2. For the juice tested, provide the following information:
   (a) Method of packaging
   (b) Storage conditions
   (c) Preservation method

3. Why were lower dilutions used for plating in this exercise than those used in the total plate count exercise in Chapter 2?

4. Fill out Table 3.4.

5. Use the yeast and mold counts in the juice you analyzed to compare the three selective media. If you were in charge of the microbiology laboratory in a company producing fruit juices, which medium would you recommend?

6. Use average class results to compare the two juices analyzed for each medium.

7. How can you confirm that a colony on acidified PDA is a yeast, not a bacterium? Explain.

8. In some commercial juice products, count of yeast and mold colonies is low at 10^0 dilution, rises at 10^{-1}, and then drops again at 10^{-2} (e.g., 50, 190, and 20 at 10^0, 10^{-1}, and 10^{-2}, respectively). Provide an explanation for this phenomenon other than "experimental error." How would you modify the experimental procedure to avoid this ambiguity?

9. Describe the colony morphology and draw the cellular morphology (Table 3.5).

10. Did the colony and morphological observations from the analyzed product resemble any of the demonstrated yeasts and molds? Comment.

11. Suggest the genus of the most common mold and that of the most common yeast in your sample.

12. Why are pathogens such as *Listeria monocytogenes* more likely isolated from blue or Camembert cheese than from Cheddar cheese?

13. Two foods (A and B) were analyzed for yeasts and molds and both were found to contain ~10^7 CFU/g.
    (a) What is food A, knowing that most consumers would not mind consuming this food?
    (b) What is food B, knowing that most consumers would refuse to consume this food?

CHAPTER 4

COLIFORM COUNT IN FOOD

PRESUMPTIVE TESTING; USING THE MOST PROBABLE NUMBER TECHNIQUE

INTRODUCTION

Coliform bacteria, which will be referred to as "coliforms," are aerobic or facultative anaerobic, gram-negative, non-spore-forming, rod-shaped bacteria that ferment lactose, with acid and gas production, within 48 hr at 35°C. Members of this group resemble *Escherichia coli* in some biochemical reactions and colony morphology. They are differentiated from other groups of microorganisms by their ability to grow in media containing bile salts (or equivalent selective agents) and also to use lactose as a carbon source with the production of acid and gas in ≤48 hr at 35°C. Notice that coliforms ferment lactose, regardless of the presence or absence of bile salts. These salts only act as selective agents against nonenteric microorganisms. Coliforms include members of at least three genera: *Escherichia*, *Klebsiella*, and *Enterobacter*. Since coliforms are common inhabitants of the intestinal tract, their presence in food may indicate fecal contamination. Therefore, coliforms are known as "indicator" microorganisms. One should be aware, however, that coliforms in food could originate from either a fecal or nonfecal origin. Additionally, presence of a large number of coliforms in food may result from growth of a small nonfecal inoculum. Thus, coliform counts in food should be cautiously interpreted.

There are three levels (phases) of testing for coliforms in water and food (Table 4.1 and Figs. 4.1 and 4.2).

(a) *Presumptive Test*   This test determines the *presumptive coliform count* since colonies with resemblance to coliforms (e.g., those producing acid or gases from lactose) are counted. If the presumptive coliform count is low, the analyst may consider the food acceptable with respect to this test and may

---

Food Microbiology   By Ahmed E. Yousef and Carolyn Carlstrom
ISBN 0-471-39105-0   Copyright © 2003 by John Wiley & Sons, Inc.

TABLE 4.1. Coliform Testing Procedures Used in This Exercise

Test	Medium	Positive Result	Result Interpretation
Presumptive	LST[a] broth	Gas in Durham tubes	Presumptive coliform count: MPN [b]/ml or MPN/g food
	VRB[c] agar	Purple-red colonies surrounded with a precipitate	Presumptive coliform count: CFU/ml or CFU/g food
Confirmatory	BGLB[d] broth	Gas in Durham tubes	Confirmed coliform count: MPN/ml or MPN/g food or CFU/ml or CFU/g food
Completed	EMB[e] broth	Black or dark center colonies (lactose fermenters)	Completed coliform count: CFU/ml or CFU/g food
		Black colonies with green sheen	Probably *E. coli*

[a] Lauryl sulfate tryptose.
[b] Most probable number.
[c] Violet red bile.
[d] Brilliant green lactose bile.
[e] Eosin methylene blue.

choose to not carry out additional testing. In case of high presumptive coliform count, the food processor may be alarmed by these results, and the analysts may opt to proceed to the second testing level (i.e., confirmation).

(b) *Confirmatory Test* This test is carried out to confirm the count obtained by presumptive testing. Confirmation is accomplished when the presumptive coliforms are analyzed for additional properties and the results are positive. For example, if the presumptive test was based on detection of gas production by lactose fermenters, confirmatory testing may be done to detect acid production from lactose under more selective conditions. Confirmation of the coliform count in food may trigger the third level of testing (i.e., a completed test).

(c) *Completed Test* When this phase of testing is carried out, at least 10% of positive confirmed tubes should be tested. The completed testing is done for different purposes. Completed testing may be designed to provide evidence that the coliforms found in food are of fecal or nonfecal origin or to prove that *E. coli* is represented in the confirmed coliform count.

OBJECTIVES

1. Determine coliform count in selected foods.
2. Become familiar with the most probable number technique.
3. Learn about presumptive, confirmatory, and completed testing phases.
4. Compare different methods of coliform testing.

PROCEDURE OVERVIEW 63

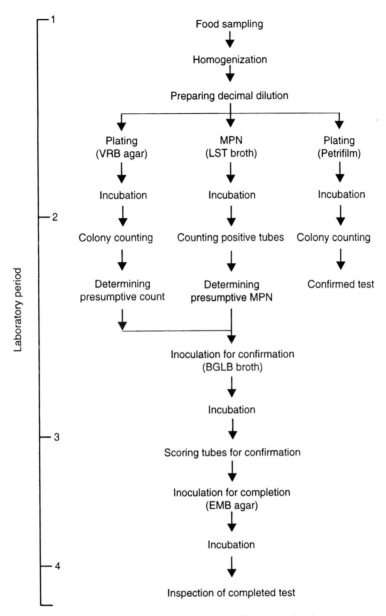

Fig. 4.1. Stages of testing for coliforms in food.

PROCEDURE OVERVIEW

This exercise gives students a chance to test raw foods and estimate their coliform load. There are many media and procedures available for counting coliforms. The choice of procedure may vary with the food tested or the information sought by the analyst. Presumptive testing will be carried out using the most probable number (MPN) techniques and a selective differential agar, violet red bile agar. This will be

64 COLIFORM COUNT IN FOOD

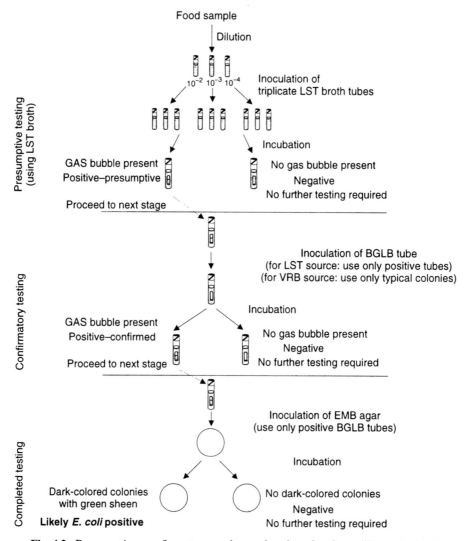

Fig. 4.2. Presumptive, confirmatory, and completed testing for coliforms in food.

followed by limited confirmation and completion testing. Additionally, a commercial dehydrated medium on film (Petrifilms) will be used as a single-step system in lieu of the presumptive and confirmation testing. It is advisable that a copy of Fig. 4.1 is annotated with essential experimental details and procedure modifications and used during the laboratory period as a guide.

An overview of the procedure of testing for coliforms in food, as applied in this exercise, is shown in Figs. 4.1 and 4.2 and Table 4.1. In the first and second laboratory periods, presumptive counts of coliforms in food will be determined using lauryl sulfate tryptose (LST) broth and violet red bile (VRB) agar. Testing the food sample using LST broth allows detection of the microorganisms that produce gas from lactose, whereas VRB agar permits detection of acid production by lactose fermenters. If positive LST broth tubes (i.e., with gas in Durham tubes) and typical colonies on VRB agar are observed, presence of coliforms in food is presumed and the resulting counts are considered presumptive counts. The positive tubes or typical colonies may be confirmed by additional testing. However, if no positive results are detected at this level, coliforms are considered absent and no further testing is required. Confirmation in this exercise will be done in the second and third laboratory periods. Positive LST broth tubes and typical colonies from VRB agar will be tested in brilliant green lactose bile (BGLB) broth for confirmation. If all BGLB broth tubes receiving inocula from the positive LST broth tubes produce positive results, this confirms the count determined by LST broth tubes. Similarly, if the selected coliform colonies from VRB agar produce positive BGLB tubes, then the presumptive coliform count obtained from VRB agar is now a confirmed count. During the third laboratory period, samples from the confirmed test (positive BGLB broth tubes) will be streaked onto eosin methylene blue (EMB) agar for completing the test, and results will be read during the fourth period.

When using the *E. coli*/coliform Petrifilm, colonies that produce gas bubbles and acid (red colonies) are considered confirmed coliforms. This count actually represents the non–*E. coli* coliforms. When *E. coli* is present, the colonies appear blue (see media description for explanation) and are surrounded by gas bubbles. Combining these two counts produces the equivalent of a confirmed coliform count obtained by other testing procedures.

Most Probable Number Technique

The MPN technique, also known as the multiple-tube technique, will be used in this laboratory exercise to determine the presumptive coliform count. Confirmation of the MPN will be attempted at the subsequent phase of testing. To analyze the food using the MPN test, the sample is decimally diluted in saline solution or peptone water, as illustrated in Chapter 1. Measured portions (e.g., 1 ml) of the sample or its dilutions are transferred into multiple tubes (three in this exercise) of selective–differential medium such as LST broth. The triplicate LST broth tubes are directly incubated at 35°C for 48 hr, and the tubes are inspected for reactions specific to coliforms. In this case, production of gas from fermentation of lactose, collected in inverted Durham tubes, presumptively indicates coliforms. Number of positive tubes in three successive dilutions is converted into a MPN/ml or MPN/g food sample using standard MPN tables (Table 4.2). The MPN is based on probability formulas, and it is an estimate of the mean density of coliforms in the sample. In addition to MPN tables, there are equations and commercially available computer programs for calculating MPN/ml or MPN/g.

In this exercise, triplicate tubes at three successive dilutions are prepared for determining the MPN. Larger number of tubes (e.g., 5 or 10) may be used for increased precision; this also increases the size of the sample analyzed and thus improves the detection limit of the method. The most satisfactory information is obtained when the multiple tubes receiving the highest dilution (i.e., the smallest sample inoculum) show no gas in all or most of the tubes while those receiving the lowest dilution show gas in some or all the tubes. For proper estimation of coliform density in the food, the sample should be homogenized adequately and mixed well before transferring the portions into the MPN tubes.

TABLE 4.2. Most Probable Number (MPN) Estimates[a] for Three Fermentation Tubes per Dilution and Dilutions Representing 0.1-, 0.01-, and 0.001-g Samples

Number of Positive Tubes/3 Tubes				Number of Positive Tubes/3 Tubes			
0.1 g	0.01 g	0.001 g	MPN/g	0.1 g	0.01 g	0.001 g	MPN/g
0	0	0	<3	3	0	0	23
0	1	0	3	3	0	1	38
1	0	0	3.6	3	1	0	43
1	0	1	7.2	3	1	1	75
1	1	0	7.4	3	2	0	93
1	2	0	11	3	2	1	150
2	0	0	9.2	3	2	2	210
2	0	1	14	3	3	0	240
2	1	0	15	3	3	1	460
2	1	1	20	3	3	2	1100
2	2	0	21	3	3	3	>1100

[a] Confidence intervals are not included to simplify the table.

MEDIA

Brilliant Green Lactose Bile Broth

This is a liquid selective–differential medium used to confirm the presence of coliforms in a food, especially in a water sample. In this laboratory exercise, BGLB broth is used in the confirmatory step. Oxgall (bile) and brilliant green inhibit gram-positives and many noncoliform gram-negative organisms. Fermentation of the lactose in the medium by coliforms will result in the production of carbon dioxide gas which is trapped by the inverted Durham tube.

Escherichia coli/coliform Petrifilms

Petrifilms are commercially available preprepared media on films instead of in plates. *Escherichia coli*/coliform Petrifilms use VRB gel medium. The medium on this film contains two dyes, neutral red, which aids in overall enumeration and has similar function as in the standard VRB agar, and an indicator of glucuronidase activity. Dye reaction results in a blue precipitate around colonies which produce β-glucuronidase, such as *E. coli*. By inclusion of this dye, the medium becomes a single-step process for the detection and enumeration of *E. coli*. For more information about *E. coli*/coliform Petrifilms, see the 3M website listed in the References.

Eosin Methylene Blue Agar

This is a selective–differential medium used for the detection and differentiation of coliforms. At acidic pH values, eosin and methylene blue combine to form a precipitate, allowing differentiation of colonies that can ferment lactose from those that cannot. These two dyes also act as selective agents, although gram-positive organisms and yeast may grow well enough to develop into pinpoint colonies. Lactose-fermenting colonies have a dark coloration. Non-lactose-fermenting colonies will appear colorless to pale straw colored. Sucrose-fermenting colonies, such as those of *Enterobacter* spp., appear pink in color. In addition, organisms such as *E. coli* that use the mixed-acid pathway may also be differentiated as these colonies have a characteristic green sheen.

Lauryl Sulfate Tryptose Broth

This medium is also known as lauryl tryptose broth, and it is considered a selective–differential medium. It is primarily used for detection of coliform bacteria in water. In this laboratory, LST broth is used as a part of an MPN determination. The inclusion of lactose in the medium allows for the detection of lactose fermentation—one of the hallmarks of coliforms. Sodium lauryl sulfate is the selective agent, that is, selecting against noncoliforms. Tryptose improves growth rate and sodium chloride provides osmotic balance, making the medium suitable for cell growth and formation of gas. Addition of inverted Durham tubes permits observation of CO_2 formation as a result of lactose use.

Violet Red Bile Agar

This is a selective–differential medium that is especially useful in the presumptive stage of coliform enumeration in food samples. Bile salts and crystal violet in the medium select against gram-positive and nonenteric organisms. The combination of lactose and neutral red results in the alteration of the medium color by organisms that ferment the lactose and produce acid end products. The lactose-fermenting colonies (such as coliforms) will be purple red. Nonlactose fermenters will produce basic end products resulting in a straw to yellow-colored medium. In addition, coliforms such as *E. coli* will precipitate bile acids from bile salts in the medium. This results in a halo of precipitate surrounding the coliform colonies. Agar overlay limits the spread of colonies. This medium is not recommended for testing food in which coliforms have been stressed or injured.

Incubation of the VRB agar plates for more than 24 hr may produce erroneous results; some slow lactose fermenters may grow enough to produce false-positive results. Additionally, tissues from vegetable samples tend to decolorize the neutral red indicator, especially when the incubation period is inadvertently extended. If the class does not meet daily, it is suggested that inoculated plates are refrigerated and then incubated 24 hr before the class meets.

ORGANIZATION

This exercise will cover four laboratory periods as shown in Fig. 4.1. Students will work in groups of two. Each group will sample one of the two foods provided. A fresh vegetable (e.g., spinach or bean sprouts) and raw meat (e.g., ground beef) will be tested. The proposed dilution scheme (Fig. 4.2) is based on assumed coliform counts in ground meat and fresh vegetables in the U.S. market. Deviations from this scheme are required if different counts are expected or alternative foods are tested.

REFERENCES

3M. 2002. *3M™ Petrifilm™ E. coli/Coliform Count Plates.* Available: http://www.3m.com/microbiology/home/products/petrifilm/petriprod/ecoli/overview.html.

American Public Health Association (APHA). 1992. *Standard Methods for the Examination of Water and Wastewater*, 18th ed. APHA, Washington, DC.

Difco Laboratories. 1998. *Difco Manual.* Difco Laboratories, Division of Becton Dickinson Co., Sparks, MD.

Doyle, M. P. 1996. Fecal Coliforms in Tea: What Is the Problem? *Food. Technol.* 50(10):104.

Jay, J. M. 2000. *Modern Food Microbiology*, 6th ed. Aspen Publishers, Gaithersburg, MD.

Kornacki, J. L., and J. L. Johnson. 2001. *Enterobacteriaceae*, Coliforms and *Escherichia coli* as Quality and Safety Indicators. In F. P. Downes and K. Ito (Ed.), *Compendium of Methods for the Microbiological Examination of Foods*, 4th ed. (pp. 69–82). American Public Health Association, Washington, DC.

Period 1 Presumptive Coliform Testing

Presumptive testing for coliforms in food will be done using the MPN technique and by counting on VRB agar medium (Table 4.1 and Fig. 4.2). Additionally, enumeration on *E. coli*/coliform Petrifilm will be run in parallel.

MATERIALS AND EQUIPMENT

Per Pair of Students

- Food sample
- Positive control (*E. coli*, diluted to 10^{-7})
- Negative control (*Proteus vulgaris*, diluted to 10^{-7})
- Peptone water (diluent): one 99-ml bottle; three 9-ml tubes
- Eight 10-ml tubes melted VRB agar (six for sample, one for positive control, one for negative control)
- Eight 5-ml tubes melted VRB agar overlay
- Eight sterile, empty Petri plates
- Ten 9.0-ml LST broth tubes (nine for the 3×3 MPN method, one for positive control)
- Seven *E. coli*/coliform Petrifilms (six for sample, one for positive control)

Class Shared

- Scale for weighing food samples (e.g., a top-loading balance with 500 g capacity)
- Incubator, set at 35°C
- Waterbath, set at 50°C

PROCEDURE

Each group will test one of the foods assigned by the laboratory instructor. A raw vegetable (e.g., spinach) and raw meat will be tested in this laboratory exercise.

Dilutions

1. Weigh 11 g of food in sterile Petri plate and transfer into a stomacher bag. The vegetable is cut and sampled using a sterile metal salad tongs and the meat is mixed throughly and a sample is taken using a sterile spatula.
2. Add 99 ml peptone water, then stomach for 2 min. The stomached sample represents the 10^{-1} dilution.
3. Prepare the following additional dilutions using the 9-ml peptone water tubes: $10^{-2}, 10^{-3}, 10^{-4}$.

MPN Method

1. Label nine tubes of LST broth. The LST tubes should contain inverted Durham tubes to detect gas production.
2. Inoculate three (triplicate) tubes with 1 ml for each of the three highest dilutions (10^{-2}, 10^{-3}, and 10^{-4}, a total of nine tubes).
3. Run a positive control by inoculating a LST broth tube with 1 ml of the *E. coli* culture.
4. Incubate all tubes at 35°C for 48 hr.

Plating with VRB Agar

(*Caution*: Do not take more than two tubes of VRB agar from the water bath at a time to prevent solidification before pouring in plates.)

Food Sample

1. Label duplicate plates at each dilution (total of six plates). Label plates also for the positive and negative control.
2. Transfer 1 ml of each of the 10^{-2}, 10^{-3}, and 10^{-4} dilutions to the appropriately labeled empty plates; two plates will receive samples from each dilution.
3. Pour the contents (10 ml) of one of the molten VRB agar tubes (~48°C) into the plate and swirl gently to mix the sample into the medium.
4. After agar solidifies, overlay each plate with 5 ml molten VRB agar.
5. After the overlay solidifies, invert plates and incubate at 35°C for 24 hr.

Positive Control

1. Transfer 1 ml of the 10^{-7} dilution of *E. coli* to the appropriately labeled empty plate.
2. Pour the contents (10 ml) of one of the VRB tubes into the plate and swirl gently to mix the sample into the medium.
3. After agar solidifies, overlay the plate with 5 ml molten VRB agar.
4. Incubate the plate at 35°C for 24 hr.

Negative Control

1. Transfer 1 ml of the 10^{-7} dilution of *P. vulgaris* to the appropriately labeled empty plate.
2. Pour the contents (10 ml) of one of the VRB tubes into the plate and swirl gently to mix the sample into the medium.
3. After agar solidifies, overlay the plate with 5 ml molten VRB agar.
4. Incubate the plate at 35°C for 24 hr.

Plating on Petrifilms

Food Sample

1. Use the last three (most dilute) of the dilutions made for VRB agar (i.e., 10^{-2}, 10^{-3}, and 10^{-4}) for plating. Label duplicate films at each dilution (total of six plates). Also, label one Petrifilm for *E. coli* control.

2. Inoculate two Petrifilms at each of the dilutions following these instructions:
    a. Lift the cover film.
    b. Transfer 1 ml of the inoculum to the surface of the labeled Petrifilm. Keeping the pipette vertical should help prevent undesired spread of the inoculum.
    c. Carefully roll the film down to avoid entrapping air bubbles.
    d. Place the spreader on top of the film over the inoculum.
    e. Gently apply pressure on the spreader to uniformly distribute the inoculum over the entire circular area ($20\,cm^2$).
    f. Lift the spreader and wait for 1 min for gel to solidify.
3. Incubate at 35°C for 24 hr.

Positive Control
1. Inoculate one Petrifilm with 1 ml of the dilute *E. coli* culture using the technique described for the food sample.
2. Incubate film at 35°C for 24 hr.

Period 2 Presumptive Testing (*Continued*) and Confirmatory Testing

During this laboratory period, results of presumptive testing will be recorded. Selected presumptive-positive LST broth tubes and colonies from VRB agar plates will be subjected to confirmatory testing (Table 4.1, Fig. 4.2).

MATERIALS AND EQUIPMENT

Per Pair of Students

- Incubated VRB agar plates
- Incubated *E. coli*/coliform Petrifilms
- Incubated LST broth tubes
- Eight tubes BGLB broth (six for subculturing LST broth and VRB agar and two for the corresponding positive controls)

Class Shared

- Incubator, set at 35°C
- Colony counters

PROCEDURE

Scoring LST Broth Tubes and Calculating the Presumptive MPN

1. Observe each of the LST tubes. Gently shake the tube to help release the gas from the medium. If there is a gas bubble trapped in the Durham tube or effervescence is produced when the tube is gently shaken, score the tube positive (+). When no gas is produced, the tube is scored negative (−). The positive control should have produced gas.
2. Record results in Table 4.3.
3. Determine the total number of positive tubes at each dilution.
4. Using Table 4.2, find the row with the corresponding positive values. For example, if there are three positives for the least dilute, two for the middle, and none for the most dilute, use the seventh row from the bottom in the second set of columns.
5. Read across the row and find the value under the column MPN/g. In this example, the value is 93.
6. Determine the dilution factor multiplier. Table 4.2 is for tubes with dilutions of 10^{-1}, 10^{-2}, and 10^{-3}. The MPN/g value from the table must be converted to the dilutions used in this experiment. For example, if the lowest dilution analyzed was 10^{-3}, there is a factor of 10^{-2} ($10^{-3}/10^{-1}$) between the table and the sample. Therefore, multiply the table values by $1/10^{-2}$ to convert the dilution. Again using the same example, the MPN value would be 9.3×10^3 MPN/g ($93/10^{-2}$ MPN/g).

TABLE 4.3. *(Write a descriptive title, including what was determined, food and methods used)*

Dilution Factor	MPN Tube Result			VRB Agar		E. coli / Coliform Petrifilm	
	1	2	3	Plate 1	Plate 2	Plate 1	Plate 2
10^{-2}							
10^{-3}							
10^{-4}							
Presumptive coliform count			MPN/g		CFU/g	NA	
Confirmed coliform count	No. positive BGLB broth tubes =			No. positive BGLB broth tubes =			
			MPN/g		CFU/g		CFU/g

(Add at least three footnotes, including those that may be needed to explain irregularities in results).

7. Record the presumptive MPN/g in Table 4.3.
8. If none of the incubated LST tubes produced gas in Durham tubes, then the testing is completed. Report the MPN/g food based on the reading from Table 4.2 and the additional calculations.

Confirming the MPN Results Using BGLB Broth

1. Select three presumptive-positive LST broth tubes for confirmation.
2. Label three BGLB tubes to match the selected LST broth tubes (i.e., dilution to be subcultured and tube number).
3. Mix the contents of the LST broth tubes by rapidly rolling each in between your hands prior to subculturing. Do not vortex since Durham tubes are inside.
4. Using a loop, subculture the three presumptive-positive LST tubes into the three labeled BGLB broth tubes (ideally, all presumptive-positive tubes would be subcultured).
5. Mix the contents of the BGLB broth tubes as described earlier.
6. Repeat using the positive control LST broth tube.
7. Incubate at 35°C for 24 hr.

Scoring the VRB Agar Plates and Calculating the Presumptive CFU/g

1. Observe the purple-red appearance of the colonies on the positive-control plate (*E. coli*). These colonies are 0.5 mm in diameter or larger and should be surrounded by a halo of precipitated bile acids.
2. Observe the appearance of the colonies on the negative-control plate (*P. vulgaris*). This microorganism forms pinpoint colonies with no halos around them. Note how the color of the medium has been changed. This is due to the non-lactose-fermenting organism using peptone as a carbon/energy source. Use of peptone results in a basic end product (ammonia). At alkaline pH, neutral red in VRB agar medium changes to a pale yellow color.
3. Count the presumptive coliform colonies on the sample's plates by matching the purple-red colony color and the precipitate to that of the positive control.
4. Record this information in Table 4.3. Keep in mind that space has been provided to include all the results but that calculations should be based only on countable plates.
5. Determine the coliform CFU/g food.
6. If none of the incubated VRB agar plates contained presumptive coliform colonies, then the testing is completed. Report the CFU/g food, based on the counting rules for plates with no colonies (See Chapter 1).

(See the precautionary notes, under "Media," about the limitations resulting from extended incubation of VRB agar plates.)

Confirming the VRB Agar Results Using BGLB Broth

1. Select three well-isolated presumptive coliform colonies from the VRB agar plates for confirmation.
2. Label three BGLB tubes to match the plates of selected colonies (i.e., dilution to be subcultured and plate number).
3. Using a loop, subculture the three presumptive coliform colonies from the VRB agar plates into the BGLB broth tubes. Since an overlay was used, some agar has to be removed with the loop to expose the colony and allow the transfer. Ideally, a larger number of presumptive coliform colonies are subcultured for an accurate confirmation.
4. Mix the contents of the BGLB broth tubes as described earlier.
5. Repeat using the positive control on VRB agar plate.
6. Incubate at 35°C for 24 hr.

E. coli/Coliform Petrifilms

1. Observe the blue color of the *E. coli* colonies on the control Petrifilm. Numerous small blue pinpoints with gas bubbles should be present.

2. Count all colonies producing gas bubbles whether they are red or blue in color; these are considered coliforms. Non–*E. coli* coliforms produce red colonies with gas, while *E. coli*, which is also a coliform, produces blue colonies with gas bubbles.
3. Record this information in Table 4.3. Keep in mind that space has been provided to include all the results but that calculations should be based only on countable plates.
4. Determine the coliform CFU/g food. Notice the added advantages of *E. coli*/coliform Petrifilms: They allow counting the coliforms and *E. coli* separately and results are considered confirmed. Record the confirmed coliform CFU/g in Table 4.3.

Period 3 Confirmatory Testing (*Continued*) and Completed Testing

During this laboratory period, results of confirmatory testing will be recorded. Selected confirmed results will be subjected to completed testing (Table 4.1 and Fig. 4.2).

MATERIALS AND EQUIPMENT

Per Pair of Students

- Incubated BGLB broth tubes.
- Seven plates, EMB agar (three for BGLB broth–LST source, three for BGLB broth–VRB source, one for the positive control)

Class Shared

- Incubator, set at 35°C

PROCEDURE

Coliform Confirmation: BGLB Tubes

LST Broth Source

1. Examine the BGLB broth tubes for entrapped gas in Durham tubes. These are the tubes that were subcultured from the positive LST broth tubes.
2. Record tubes with gas as positive (Table 4.3).
3. If all three BGLB broth tubes are positive, this confirms the LST broth results. Therefore, the previous presumptive MPN/g from LST broth tubes is now confirmed.
4. If one or two BGLB broth tubes are positive, calculate coliform MPN/g food based on the proportion of confirmed gassing in BGLB broth tubes.
5. If all BGLB broth tubes are negative, then the presumptive coliform MPN is not confirmed and the coliform group is considered absent.

VRB Agar Source

1. Examine BGLB broth for entrapped gas in Durham tubes. These are the tubes that were subcultured from the presumptive coliform colonies on the VRB agar plates.
2. Record tubes with gas as positive (Table 4.3).
3. If all three BGLB broth tubes are positive, this confirms the VRB agar results. Therefore, the previous presumptive CFU/g from the VRB agar plates becomes confirmed.
4. If one or two BGLB broth tubes are positive, calculate coliform CFU/g food based on the proportion of confirmed gassing in BGLB broth tubes.

5. If all BGLB broth tubes are negative, then the presumptive coliform CFU/g is not confirmed and the coliform group is considered absent.

Completed Test Using EMB

1. Mix gently the contents of the positive BGLB broth tubes as indicated earlier.
2. Using a loop, subculture the positive BGLB broth tubes onto EMB agar plates by three-phase streaking.
3. Repeat the former steps using the positive control.
4. Incubate the plates at 35°C for 24 hr.

Period 4 Completed Testing Results

MATERIALS AND EQUIPMENT

Per Pair of Students
- Incubated EMB agar plates
- Colony counter

PROCEDURE

This is a qualitative test that serves as an additional positive confirmation of coliforms in food.

1. Inspect EMB agar plates.
2. Observe colonies with dark centers with or without green sheen; these are typical lactose-fermenting coliforms. *Escherichia coli* typically produces green sheen.
3. Observe pink mucoid colonies, which indicates sucrose-fermenting coliforms such as *Enterobacter* spp.
8. Noncoliform bacteria, such as *Enterococcus faecalis*, appear colorless.

PROBLEMS

1. What processing conditions or methods for the food sample analyzed in the laboratory may cause high coliform counts?

2. Why were these foods chosen for testing? What other foods would have been good choices for testing in this experiment?

3. For the food tested in this laboratory by the group, list the following information:

    (a) Method of packaging
    (b) Storage method
    (c) Sell-by date (if available)

    What additional information about the food analyzed may have helped in interpreting the results?

4. List the experimental variables that were tested in this laboratory exercise.

5. Record and analyze the data as follows:

    (a) Fill in Table 4.3 using the group raw data with a descriptive title.
    (b) Add at least three footnotes to Table 4.3; include abbreviations or places where more information should be added.
    (c) Report the confirmed coliform counts for the two food samples when MPN, plating (VRB agar), and Petrifilm techniques were used.
    (d) Construct a table for the average class data, reporting confirmed coliform counts for the two food samples, when MPN, plating (VRB agar), and Petrifilm techniques were used.

6. Explain the purpose of a positive control. Why was *E. coli* used as the positive-control bacterium in this experiment? What is the purpose of a negative control?

7. Was *E. coli* represented in the coliform counts obtained? Explain. Which test allows counting *E. coli* specifically?

8. Does gas production in the LST tube of the MPN confirm the presence of coliforms? If not, how would this test have been modified to include confirmation?

9. Compare coliform counts from the three different media (LST broth, VRB agar, *E. coli*/coliform Petrifilms). Describe why using these different media may yield different results.

10. Compare counts that the group obtained with the class average for the same type of food.

11. Compare counts between the foods that the class tested. Explain differences in the counts.

12. Where did most of the *E. coli* on raw vegetables probably originate?

13. Based on the results obtained, comment on the microbiological quality of the food analyzed. Remember that the limited data gathered during this labo-

ratory exercise may not be sufficient to judge the suitability of this food for consumption.

14. Explain why the enterohemorrhagic *E. coli* (*E. coli* O157:H7) is not detected by any of the tests used in this laboratory exercise.

15. In 1996, a large count of fecal coliforms was detected in restaurant ice tea [see Doyle (1996) in the References]. What were the likely sources of this contamination? Would this tea be considered unsafe for drinking?

16. In an article published in the *National Provisioner* (April 1997), the author indicated that some food microbiologists consider the Enterobacteriaceae test of more value than the coliform test. Do you agree with viewpoint? Why?

CHAPTER 5

MESOPHILIC AEROBIC AND ANAEROBIC SPORES

USING HEAT AS A SELECTIVE FACTOR;
AEROBIC VERSUS ANAEROBIC INCUBATION;
EXAMINATION OF SPORES

INTRODUCTION

Spore-forming bacteria of significance in food include three genera, *Bacillus*, *Alicyclobacillus*, and *Clostridium* (Table 5.1). While *Clostridium* spp. are obligate or aerotolerant anaerobic bacteria, *Bacillus* and *Alicyclobacillus* are aerobes or occasionally facultative anaerobes. *Alicyclobacillus* spp. have similarity to *Bacillus* spp., but they are known to be acidophiles. These genera form gram-positive rods and produce spores. Dormant bacterial spores are typically refractile (shiny) when seen by phase-contrast microscopy. *Clostridium* spp. are catalase negative whereas most *Bacillus* spp. are catalase positive.

Bacillus and *Clostridium* spp. are commonly found in soil and thus are considered natural contaminants of most foods, particularly vegetables. Some *Clostridium* spp. are present in the intestinal tract of animals and appear as contaminants in food of animal origin. Because of their widespread nature and tolerance to dryness and other lethal factors, bacterial spores are commonly found in spices, cereal grains, dried fruits, flour, starches, and dried milk.

Spore-forming bacteria are common causes of food spoilage (Table 5.1). *Bacillus* spp. cause rapid deterioration of pasteurized milk, softening and stickiness of the center of bread loaf (ropy bread spoilage), and flat sour spoilage (acid but no gas production) in containers of thermally processed low-acid (pH > 4.6) food. *Alicyclobacillus* spp. are particularly problematic in pasteurized fruit juices and occasionally cause spoilage of these products. *Clostridium* spp. produce gas in cheese blocks at their late stage of ripening (late gassiness defect) and spoil canned products such as sweet corn and spaghetti with tomato sauce. When some *Bacillus* and *Clostridium* spp. grow in food, they produce toxins that cause foodborne diseases.

Food Microbiology   By Ahmed E. Yousef and Carolyn Carlstrom
ISBN 0-471-39105-0   Copyright © 2003 by John Wiley & Sons, Inc.

TABLE 5.1. Spore-Forming Bacteria of Importance in Food

Bacteria	Properties	Spore Characteristics	Importance in Foods
	MESOPHILIC AEROBES		
Bacillus anthracis	Catalase positive; facultative anaerobe; large cell; no hemolysis; nonmotile; hydrolyze gelatin, starch, casein; optimum growth at 30–40°C; growth in 7% NaCl	Ellipsoidal, central to subterminal; no swelling of sporangium	Foodborne infection; linked to ingestion of undercooked infected meat
Bacillus cereus	Catalase positive; facultative anaerobe; large cell; hemolysis; hydrolyze gelatin, starch; growth at 5–50°C (optimum 28–35°C); growth at pH 4.9–9.3	Ellipsoidal, central to subterminal; no swelling of sporangium; $D_{100°C}*$ = 3–200 min	Food intoxication; disease outbreaks linked with starchy foods (cooked rice, pasta), cooked meats, vegetables, soups, salads, puddings
Bacillus licheniformis	Catalase positive; facultative anaerobe; growth at 30–55°C growth in 7% NaCl	Oval, central to subterminal; no swelling of sporangium; $D_{100°C}$ = 13.5 min	Spoilage of cooked meat (softening, discoloration); opportunistic pathogen
Bacillus megaterium	Catalase positive; obligate aerobe; large cell; growth at 5–40°C	Oval, round, elongated central to subterminal; no swelling of sporangium; $D_{100°C}$ = 1 min	Spoilage: coagulation of canned evaporated milk
Bacillus polymyxa	Catalase positive; facultative anaerobe; growth at 5–40°C; growth at pH 5	Oval endospore; swelling of sporangium; $D_{100°C}$ = 0.1–0.5 min	**Spoilage** of canned vegetables (swelling)
Bacillus subtilis	Catalase positive; obligate aerobe; growth at 10–50°C; growth at pH 5.5–8.5	Oval, central to subterminal; no swelling of sporangium; $D_{100°C}$ = 7–70 min	Spoilage: "ropy" bakery product, softening of pickles; opportunistic pathogen

THERMOPHILIC AEROBES

Organism	Characteristics	Spore	Spoilage
Alicyclobacillus spp. *A. acidoterrestris* *A. acidocaldarius* *A. cycloheptanicus* *A. herbarius* *A. hesperdium*	Catalase positive; hydrolyze starch; no growth in 5% NaCl; acidophile: growth at pH 2.5–5.5 (optimum 3.5–4.0); optimum growth at 45–50°C	Oval, subterminal to terminal; swelling of sporangium; $D_{95°C} = 2.2–8.7$ min	Flat sour spoilage; spoilage of fruit juices and iced tea
Bacillus coagulans	Catalase positive; facultative anaerobe; no growth in 5% NaCl; growth at 30–65°C (optimum > 50°C); aciduric: growth at pH 4.5 (optimum pH 6)	Oval endospore; swelling of sporangium; high D value: $D_{100°C} = 20–300$ min	Flat sour spoilage; spoilage of canned tomato juice and other acid foods; coagulation of canned condensed milk
Bacillus stearothermophilus	Facultative anaerobe; growth at 40–70°C (optimum > 50°C); no growth in 7% NaCl; limited tolerance to acid	Oval endospore; swelling of sporangium; high D value: $D_{100°C} = 100–1600$ min	Flat sour spoilage; spoilage of low-acid canned vegetables; coagulation of canned condensed milk

MESOPHILIC ANAEROBES

Organism	Characteristics	Spore	Spoilage
Proteolytic *Clostridium botulinum*	Catalase negative; produce toxin types A, B, F; digest gelatin, milk, meat; growth at 10–48°C (optimum 35–40°C) minimum growth pH > 4.6; no growth in 6.5% NaCl or at pH 8.5	Oval, subterminal endospore; swelling of sporangium; $D_{100°C} = 15–25$ min	Food intoxication; food implicated: home-canned foods, foods subjected to faulty commercial processing or temperature abuse, vegetables, particularly those in contact with soil
Nonproteolytic *C. botulinum*	Catalase negative; produce toxin types B, E, F; saccharolytic; digest gelatin; growth at 3–45°C (optimum 25–37°C); minimum growth pH > 5.0; no growth in 6.5% NaCl or at pH 8.5	Oval, subterminal endospore; swelling of sporangium; $D_{100°C} < 0.1$ min	Food intoxication; food implicated: fermented marine products, dried fish, vacuum-packed fish
Clostridium butyricum	Catalase negative; some strains produce toxin type E; optimum growth at 30–37°C; minimum growth >10–15°C; minimum growth pH > 4.0–5.2; no growth in 6.5% NaCl	Oval, central to subterminal; no swelling of sporangium; $D_{100°C} < 1–5$ min	Spoilage of canned tomatoes, peas, olives, cucumbers (swelling and butyric odor)

TABLE 5.1. (*Continued*)

Bacteria	Properties	Spore Characteristics	Importance in Foods
Clostridium perfringens	Catalase negative; air tolerant; rapid growth; stormy fermentation of lactose in milk; growth at 20–50°C (optimum 43 to 45°C); growth at pH 5.5–8.5; no growth in 6.5% NaCl	Large, oval, central or subterminal; swelling of sporangium; no exosporium; $D_{100°C} = 0.3$–$18\,\text{min}$	Foodborne noninvasive infection; food implicated: meat and poultry product; outbreaks usually due to temperature abuse
Clostridium sporogenes	Catalase negative; putrefactive anaerobe; growth at 25–45°C (optimum 30–40°C); good growth in 100% CO_2; growth in 6.5% NaCl or at pH 8.5	Oval, subterminal; swelling of sporangium; $D_{100°C} = 80$–$100\,\text{min}$	Spoilage of canned vegetables (swelling and putrid odor); putrefaction of cured bacon; off odor of Swiss cheese
Clostridium tyrobutyricum	Catalase negative; some strains produce toxin type E; growth at 25–45°C (optimum 30 to 37°C); no growth in 6.5% NaCl	Oval, subterminal endospore; swelling of sporangium	Gas formation ("blowing") of cheeses
THERMOPHILIC ANAEROBES			
Clostridium thermosaccharolyticum	Catalase negative; optimum growth at 55–62°C; some slowly grow at 37°C; few grow at 30°C poorly; no growth at 70°C	Oval/round, terminal; swelling of sporangium; high D value: $D_{100°C} = 400\,\text{min}$	Spoilage of canned vegetables (swelling and sour, butyric odor)
Desulfotomaculum nigrificans	Catalase negative; obligate thermophile: minimum growth temperature $\geq 43°C$, optimum 55°C; growth at pH 5.6–7.8 (optimum pH 6.8–7.3)	Oval/round, terminal to subterminal; slight swelling of sporangium; high D value: $D_{100°C} < 480\,\text{min}$, $D_{120°C} = 2.0$–$3.0\,\text{min}$	Hydrogen sulfide spoilage of low-acid canned vegetables (blackened appearance and rotten-egg odor); no swelling of can.

Table compiled by Y.-K. Chung.
* $D_{x°C}$: Time of heating at $x°C$ required to inactivate 90% of microbial population.

Bacillus cereus and *C. botulinum* are notable for causing food intoxications. When *C. perfringens* is ingested in large numbers with food, the bacterium sporulate in the intestine, a toxin is released during the sporulation process, and a noninvasive infection follows.

Presence of high counts of bacterial spores in food may indicate heavy environmental contamination, poor handling and storage, or lack of processing treatments that eliminate spores. Commercial sterilization of canned foods (e.g., retorting) is designed to eliminate pathogenic spore formers. Gamma radiation decreases spices' spore load considerably. Bacterial spores in food, however, survive processing treatments that are less severe than retorting. Faulty thermal processing also results in food containers with surviving bacterial spores. In this laboratory exercise, spores of aerobic and anaerobic mesophilic bacteria will be counted in selected foods. This exercise allows counting the spores, but not the vegetative cells, of spore-forming bacteria.

Spore Structure and Resistance

Microorganisms produce different types of spores for reproductive and protective reasons. While fungi produce both reproductive and protective types, bacterial spores are produced mainly for protection against unfavorable environments. The structure of the bacterial spore is related to its resistance. Morphologically, the bacterial spore has a distinct structure (Fig. 5.1). Many spore structures, including the exosporium and coats, have no counterparts in the vegetative cell. The outermost spore layer, the exosporium, varies significantly in size between species. Underlying the exosporium are the spore coats, which are two layers (i.e., outer and inner spore coats). The spore coats protect the spore cortex from attack by lytic enzymes. The coats also may provide an initial barrier to chemicals such as oxidizing agents. However, the coats seem to provide no protection to the spore against physical food preservation factors such as heat or irradiation. The outer coat and exosporium contain mainly spore proteins, which are unique in terms of composition. These proteins are rich in the more hydrophobic amino acids and contain a high amount of cysteine. Hydrophobicity and disulfide bridges in these outer layers may be respon-

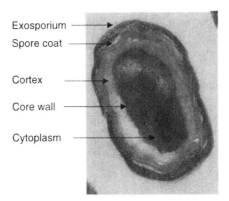

Fig. 5.1. Transmission electron micrograph of *Bacillus subtilis* spores. (Courtesy of M. Khadre.)

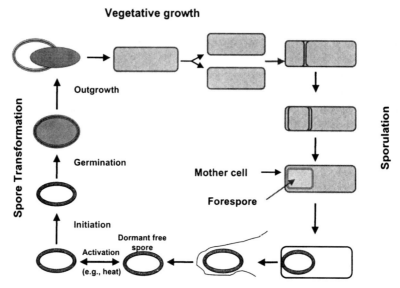

Fig. 5.2. Life cycle of a spore-forming bacterium.

sible for maintaining the dormant state of the spore. The inner spore coat (the cortex) is largely made of peptidoglycan and thus is structurally similar to the bacterial cell wall, but with several differences, including being loosely cross-linked and lacking amino acid cross-links between adjacent peptide chains. The germ cell wall and inner membrane are structurally similar to their counterparts in the vegetative cell. The innermost region, the core, contains the deoxyribonucleic acid (DNA), ribonucleic acid (RNA), ribosomes, and most enzymes as well as dipicolinic acid (DPA) and divalent cations. Low water content in the core may play a major role in spore dormancy and in spore resistance to various agents.

Spore-to-Cell Transition

The complete life cycle of a spore-forming bacterium includes multiplication of the vegetative cell by binary fission, sporulation, and spore-to-cell transformation (Fig. 5.2). The spore-to-cell transformation stage is sometimes referred to as germination, while in fact germination is only one step of this transformation process. The transition of dormant bacterial spores to fully active vegetative forms can be divided into three sequential processes: activation, germination, and outgrowth.

Activation Activation of spores or the breaking of dormancy is commonly achieved by heating spores in aqueous suspension. The temperature and duration of optimal heating vary widely among different species and even among different spore preparations of the same strain. Commonly used heat treatments fall within these ranges: 75–80°C and 15–30 min.

Germination Germination is defined as a series of degradative events triggered by specific germinants and leads to the loss of typical spore properties. Presence of germinants, such as certain amino acids and sugars, triggers the germination of activated spores. These changes are divided into two categories: (1) those events detected within the first few minutes of germination triggering, including loss of heat resistance, commitment, and dipicolinic acid (DPA) release, and (2) those events initiated at a later stage, including loss of light absorbency, cortex hydrolysis, and the onset of spore metabolism.

Outgrowth Outgrowth refers to the emergence of bacterial vegetative cells from the germinating spore. These cells are more sensitive than spores to heat and other deleterious factors.

OBJECTIVES

1. Determine the counts of spore formers in selected foods.
2. Compare the aerobic and anaerobic spore load in food.
3. Learn how to use heat as a selective agent in enumerating spores.
4. Become familiar with the morphology of bacterial endospores.

PROCEDURE OVERVIEW

The procedure used in this exercise is illustrated in Fig. 5.3. Briefly, the food is sampled, the sample is homogenized in a suitable diluent (e.g., peptone water), and dilutions are made. One milliliter of each dilution is mixed with molten agar medium in test tubes and the mixture is promptly heated in a waterbath at 80°C for 30 min. The heat treatment kills the vegetative microbial cells in the sample but not the bacterial spores, activates these spores, and enhances their ability to germinate. The heat treatment, therefore, selects for bacterial spores. Delay in heating may cause spores to begin germination and lose heat resistance; this leads to an underestimation of spore count in the sample. The heated mixtures are cooled to ~50°C and poured in Petri plates. Some plates receive additional agar medium containing a reducing agent (e.g., thioglycolate agar overlay). These plates are incubated anaerobically and used for determination of the anaerobic mesophilic spore count. The other plates do not receive agar overlay and are incubated aerobically to determine the aerobic mesophilic spore count. Plates are counted following incubation. Some plates are incubated for an additional 48 hr and sporulation of cells in selected colonies will be observed using phase-contrast microscopy and a spore-staining technique.

Unprocessed spices (e.g., organic spices) and unbleached flour are analyzed in this exercise; these food ingredients may contain considerable spore load. Adjustments of this procedure may be required if different spore loads are expected in these foods or if other foods or food ingredients are tested. When the tested product has a low spore count, greater sample size is plated to increase the test's detection limit. In this case, multiple tryptone glucose extract (TGE) agar tubes (e.g., five tubes) are prepared for each dilution, and the total of colony counts on all plates receiving a given dilution is reported. This increases the size of the food sample that is actually plated and thus increases the detection limit of the test. When testing foods with sizable spore loads, as done in this laboratory exercise, only duplicate TGE tubes are needed.

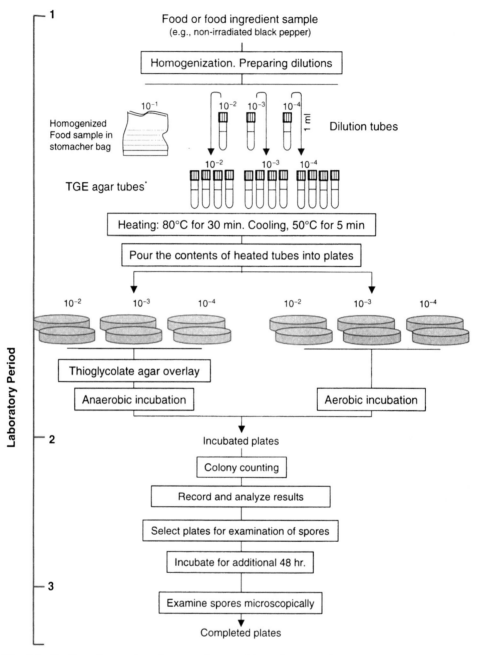

Fig. 5.3. Outline of procedure for counting aerobic and anaerobic mesophilic spore formers. Dilution tubes and TGE agar tubes are labeled with the same dilution factors since the entire contents of the latter tubes are transferred to Petri plates.

Anaerobic Incubation

Anaerobic jars are airtight containers that have oxygen removed or reduced in level and each normally holds 10–20 Petri plates. Chemical reactions that consume oxygen are initiated in a reagent packet by addition of water. The activated packet is placed in the jar, along with an anaerobic indicator, and the jar is sealed. Different reagent packets and anaerobic indicators are available commercially. Anaerobic chambers are also useful when a large number of plates need to be incubated anaerobically. In this case, the oxygen in the chamber is replaced with another gas (commonly nitrogen) from an external source (e.g., a nitrogen tank). Using a jar and candle is the oldest method to achieve anaerobic conditions for incubated plates. A candle is lit next to the Petri plate stack and both are covered with a bell-shaped glass jar. The lit candle consumes the oxygen in the confined space under the jar and the flame is autoextinguished. Anaerobic conditions are maintained during the incubation of the plates provided the setup is airtight.

MEDIA

Tryptone glucose extract agar (also known as yeast dextrose agar) is a nonselective medium. This medium has been used historically for detection and enumeration of bacteria in dairy products. Because there are numerous amino acids and carbon/energy sources in this medium, TGE agar is also useful for promoting the germination of bacterial endospores. An overlay of thioglycolate agar is applied on plates to be incubated anaerobically. Sodium thioglycolate is a reducing agent that binds with free oxygen. By adding this chemical to a medium, conditions are rendered essentially anaerobic. Use of thioglycolate agar as an overlay or as the plating medium can facilitate the growth of anaerobic spore formers, that is, *Clostridium* spp.

ORGANIZATION

Possible food ingredients to be tested include flour, starches, and spices. Organic spices and unbleached flour will be used in this exercise. One food sample will be analyzed by a pair of students. Both members of the group will cooperate to test the sample in a timely fashion.

In preparation for this exercise, the laboratory procedure should be read carefully in advance. Annotating the entries in Fig. 5.3 with appropriate experimental details is also advisable. The annotated flowchart (see Chapter 1) may be used as a guide while executing the laboratory exercise. Deviations from the listed procedure may also be included in the annotations.

REFERENCES

Banwart, G. J. 1989. *Basic Food Microbiology*, 2nd ed. Van Nostrand Reinhold, New York.

Scott, V. N., J. E. Anderson, and G. Wang. 2001. Mesophilic Anaerobic Sporeformers. In F. P. Downes and K. Ito (Eds.), *Compendium of Methods for the Microbiological Examination of Foods*, 4th ed. (pp. 229–238). American Public Health Association, Washington, DC.

Setlow, P., and E. A. Johnson. 2001. Spores and Their Significance. In M. P. Doyle, L. R. Beuchat, and T. J. Montville (Eds.), *Food Microbiology: Fundamentals and Frontiers*, 2nd ed. (pp. 33–70). American Society for Microbiology, Washington, DC.

Sneath, P. H. 1986. Endospore-Forming Gram-Positive Rods and Cocci. In P. H. Sneath, N. S. Mair, M. E. Sharpe, and J. G. Holt (Eds.), *Bergey's Manual of Systematic Bacteriology*, Vol. 2 (pp. 1104–1207). Williams & Wilkins, Baltimore, MD.

Stevenson, K. E., and W. P. Segner. 2001. Mesophilic Aerobic Sporeformers. In F. P. Downes and K. Ito (Eds.), *Compendium of Methods for the Microbiological Examination of Foods*, 4th ed. (pp. 223–228). American Public Health Association, Washington, DC.

Period 1 Sample Preparation, Plating, and Incubation

MATERIALS AND EQUIPMENT

Per Pair of Students

- Food sample
- A bottle of 99 ml peptone water
- Three 9-ml peptone water tubes
- Twelve 9-ml TGE tubes (two incubation conditions, three dilutions, two replicates)
- Six 5-ml thioglycolate agar tubes (overlay)
- Twelve empty, sterile Petri plates

Class Shared

- Scale for weighing food samples (e.g., a top-loading balance with 500 g capacity)
- Anaerobic system: Anaerobic jar, reagent packet, and anaerobic indicator
- Incubator set at 35°C
- Stomacher
- Waterbath set at 50°C
- Waterbath set at 80°C

PROCEDURE

Dilution Preparation

1. Weigh 11 g of food directly into a stomacher bag, then add 99 ml diluent.
2. Homogenize the sample–diluent mixture in a stomacher for 2 min; the homogenized sample represents the 10^{-1} dilution.
3. Label the diluent (9-ml peptone water) tubes and prepare the 10^{-2}, 10^{-3}, and 10^{-4}.

   (*Caution:* Avoid transferring food particles from the homogenized sample into dilution tubes and subsequently into the TGE agar tubes; food particles in the agar medium may mask colonies and are sometimes misinterpreted as colonies.)

4. Test dilutions 10^{-2}, 10^{-3}, and 10^{-4}; four TGE agar tubes are needed for each dilution.

   (*Caution:* Do not remove the tubes from the 50°C waterbath until ready to inoculate them. Work with no more than four tubes at a time to prevent premature solidification of the medium.)

5. Label 12 TGE agar tubes (two incubation conditions, three dilutions, two replicates). Include dilution, tube number, and group identification.
6. Transfer 1-ml aliquots of each of the 10^{-2}, 10^{-3}, and 10^{-4} dilutions into four TGE agar tubes (two for each of the aerobic and anaerobic plates).

   (*Caution:* Place inoculated tubes without delay into the 80°C waterbath.)

Heat Treatment

It is advisable for a student group to keep their tubes together in one rack and for two groups to share a test tube rack in the waterbath; this makes it easy to keep track of group samples.

1. Heat the agar tubes at 80°C for 30 min in a waterbath.
2. Agitate the tubes occasionally to improve uniformity of agar heating.
3. Cool the heated tubes in a waterbath at 50°C for ~5 min. Leave tubes in the waterbath until ready to pour to prevent premature solidification.

Plating

1. Label Petri plates with the dilutions 10^{-2}, 10^{-3}, and 10^{-4} for anaerobic incubation. A total of six plates (two for each dilution) are needed.
2. Label Petri plates with the dilutions 10^{-2}, 10^{-3}, and 10^{-4} for aerobic incubation. A total of six plates (two for each dilution) are needed.
3. Pour the contents of the heated food–agar media into the appropriately labeled Petri plates. Hold plates at room temperature until agar solidifies.
4. Plates prepared for anaerobic incubation receive an additional 4–5 ml thioglycolate agar as an overlay. After the agar solidifies, invert the plates and incubate in an anaerobic jar at 35°C for 48 hr.
5. Invert plates prepared for aerobic incubation and incubate aerobically at 35°C for 24 hr.

Period 2 Colony Counting and Calculations

MATERIALS AND EQUIPMENT

Per Pair of Students
- Incubated aerobic and anaerobic plates
- Colony counter

PROCEDURE

1. Count colonies in the aerobic and anaerobic plates using a colony counter. As a safety precaution, keep the plates shut, using tape, while counting colonies.
2. Record the colony counts in Table 5.2.
3. Calculate the spores per gram of food sample using the counting rules in Chapter 1.
4. Select two aerobically incubated plates with well-isolated colonies for additional incubation and spore inspection. Incubate the plates aerobically for an additional 48 hr at 35°C.

TABLE 5.2. Counts of Colonies on Tryptone Glucose Extract Agar Plates and Spore Count in Food After Plating Diluted Samples and Incubating Plates Under Mesophilic Aerobic and Anaerobic Conditions

Anaerobic Incubation[a]			Aerobic Incubation[b]		
Dilution Factor	Plate Numer	Colony Count	Dilution Factor	Plate Number	Colony Count
	1			1	
	2			2	
	1			1	
	2			2	
	1			1	
	2			2	

[a] Spores/g =
[b] Spores/g =

(Show calculations as a footnote.)

Period 3 Microscopic Examination of Spores

MATERIALS AND EQUIPMENT

Per Pair of Students

- Aerobic plates after extended incubation
- Microscope with phase-contrast attachments
- Microscope slides and coverslip
- Malachite green stain
- Safranin stain

Class Shared

- Boiling waterbath with a staining rack mounted over the boiling water

PROCEDURE

Phase-Contrast Microscopy

1. Select four colonies and aseptically transfer a small portion of each to a drop of water on the microscope slide.
2. Mix the cell mass with the water drop and add the coverslip.
3. Examine the slide using the phase-contrast microscope.
4. Observe the presence of free refractile (shiny) spores or endospores.
5. Estimate roughly the cell–spore ratio.
6. Draw and describe the endospores observed.

Spore Staining

1. Prepare heat-fixed smears of four colonies from the aerobic plates as explained in Chapter 1.
2. Place the slides containing the heat-fixed smears (with the smear side up) on the staining rack of the waterbath.
   (*Caution:* Handling the slides on the waterbath rack should be done carefully to avoid any injuries or burns.)
3. Add malachite green stain to the slides and stain for 5 min. If needed, add more dye to compensate for evaporation.
   (*Caution:* Wear gloves when handling the malachite green stain to avoid staining the hands and exposure to this hazardous chemical.)
4. Rinse the stain off the slide using water.
5. Add safranin to the slide and stain for 60 sec.
6. Rinse off the stain with water and blot dry.
7. Examine the stained smears using the microscope's oil immersion lens.

8. Notice the green-stained spores and the red-stained vegetative cells.
9. Observe the presence of spores within parent cells (sporgia) or in a free state.
10. Determine the spore placement in the parent cell, that is, central, subterminal, or terminal.
11. Record these observations.

PROBLEMS

1. Provide the following information about the food analyzed in this laboratory exercise:
   (a) Name and brand
   (b) Method of packaging (e.g., canned, vacuum packed, or packaged under nitrogen environment) and package type
   (c) Storage conditions (refrigeration, freezing, or shelf storage)
   (d) Product sell-by or use-by date
   (e) Preservatives (if present) or preservation method

2. Why were these foods chosen for testing for spores?

3. What processing treatments did these foods receive that would affect their microbial loads?

4. What are the experimental variables that were tested in this laboratory? Why is it important to test for these variables?

5. Were selective or differential media used in this test? If yes, what selective or differential agents were present in the medium? If not, how were mesophilic aerobic and anaerobic spores selectively enumerated in food?

6. Why was it important that the inoculated tubes be transferred immediately to the 80°C waterbath? Explain how the results would be affected if the transfer was delayed.

7. Compare the group results to the class results for the analyzed food. Include counts and comment on agreement or disagreement of numbers and possible reasons for differences.

8. Spore formation is an important stage of the life cycle of a microorganism. Why would a bacterium produce spores? Why would a mold produce spores?

9. What are the conditions most favorable for sporulation of bacteria in food and the environment? Which of these conditions apply to the food analyzed in this exercise?

10. What are the conditions conducive to spore germination? Which of these conditions apply to the experimental procedure used in this exercise?

CHAPTER 6

MICROBIOTA OF FOOD PROCESSING ENVIRONMENT

FIELD SAMPLING; DETECTION OF *Pseudomonas spp.*

INTRODUCTION

Microorganisms in the processing environment contribute to the food microbiota. A heavily contaminated environment commonly leads to poor-quality products. The hazard of a secondary contamination of food with pathogens is expected if these pathogens are commonly isolated from the factory's environment. Therefore, samples for microbiological analysis are taken more often from the processing environment than from food. These samples are commonly taken from floors, drains, and equipment surfaces, particularly those that are food contact surfaces. Refrigerators and other storage sites also should be sampled frequently.

Analysis of environmental samples allows investigators to evaluate secondary contamination sources in the processing environment. Sites with a history of frequent contamination are sampled more often than others. A prior knowledge of the processing facility and an understanding of the past and potential contamination and safety problems of this facility should aid in developing a sound sampling strategy. Results of the analysis may help food processors determine the appropriate frequency of cleaning and sanitizing or verify the efficacy of a cleaning and sanitization procedure. Careful analysis of environmental sampling results may help locate a worn-out air filter or detect a faulty equipment design. Attachment of microorganisms to surfaces and the buildup of biofilm in the processing facility are common problems that may be revealed by proper environmental sampling. Recurrence of food contamination with a particular microorganism may be traced back to a location in the processing facility where the microorganism thrives. Therefore, a good environmental sampling scheme and proper analysis of results may help define critical control points in a food processing operation. This may

---

Food Microbiology By Ahmed E. Yousef and Carolyn Carlstrom
ISBN 0-471-39105-0 Copyright © 2003 by John Wiley & Sons, Inc.

also help establish a successful hazard analysis and critical control points (HACCP) plan.

Environmental sampling may be done to count the total microbial population or some indicator microorganisms or to detect a spoilage or a pathogenic microorganism of concern. Detection of *Salmonella* spp. and *Listeria monocytogenes* in the processing environment is of particular significance. Presence of *Listeria* spp. in environmental samples from a dairy factory or a seafood processing facility should be carefully considered by the processor and corrective measures should follow. Similarly, presence of *Salmonella* spp. in the effluents of a meat processing factory or in the chicken feces of an egg production hen house should alarm the management of these establishments about the potential risk of product contamination. Presence of pathogens in environmental samples, therefore, may provide an early warning to processors and should help devise corrective measures to ensure the safety of the product and consumers.

Criteria for acceptable microbiological results from food contact surfaces depend on the food being processed in the facility. Adequate cleaning and sanitizing should decrease surface contaminants by 4–5 log units. Residual microbial population on these surfaces may not exceed 10^2 CFU per utensil or surface area sampled.

Environment Sampling Methods

The technique of sampling depends on the nature of the site, degree of contamination, and microbiological information sought. There are two categories of samples that are commonly taken from food processing environments: surface and air.

Surfaces Sampling It is crucial to examine food contact surfaces for microbial contaminants. A food contact surface could be part of an equipment, packing material, storage tank, ripening room, and others. Surfaces that do not directly contact food may also cause product contamination. These include enclosures, walls, floors, and workers' garments. Sampled surfaces could be rough or smooth, flat or with curves and corners, continuous or with cracks and crevices, and accessible or difficult to reach. Therefore, the choice of a method to sample a surface depends on surface properties.

Swab Method A sterile cotton swab is commonly used. The swab is made of a wound cotton head (~0.5 cm diameter and 2 cm long) and a 12–15-cm long wooden stick. Cotton swabs may be made and sterilized by the analyst or bought as individually wrapped sterile swabs. In addition to the swab, a sterile rinse solution is needed for surface sampling. The details of the swabbing procedures will be given throughout this exercise.

Sponge Method If the surface to be swabbed is large or a microorganism of a low incidence rate in the processing environment (e.g., *Salmonella* spp.) is sought, a sponge may be used instead of the cotton swab. A natural or synthetic sponge with ~5 × 5-cm contact surface and free from antimicrobial agents is suitable for this purpose. The sponge can be packed in a heat-resistant bag or wrapped in aluminum foil and sterilized by the analyst or purchased prepacked and presterilized from a commercial source. During sampling, the sponge is held aseptically, moistened with 10 ml rinse solution, rubbed against the surface to be sampled, and returned to a sterile plastic bag. The sample should be transferred to the laboratory under refrigeration and analyzed without delay. If the purpose of sampling is to detect pathogens, the sponge is transferred to a suitable enrichment broth and the mixture is incubated. When sampling is done to quantify environment microbiota, the sponge is mixed with 50 or 100 ml diluent and further dilutions are made. Dilutions are then plated as indicated later in this chapter.

Replicate Organism Direct Agar Contact (RODAC) Method The RODAC method may be used on easily accessible flat surfaces. In this method, plates are filled with an agar medium suitable for the microbiota analyzed; these plates may be prepared in the laboratory or purchased from a commercial source. The RODAC plates should contain enough agar medium so that the convex surface of the medium rises above the rim of the plate. At the sampling site, the agar medium in the RODAC plate is exposed to the surface being sampled. This is accomplished by pressing the plate against the sampled surface and rolling the plate while applying the pressure. The cover is replaced and the plate is incubated at a temperature and for a time appropriate to the targeted microorganism or the microbiota. After incubation, the colonies on each plate are counted and colony subculturing may follow. Since no sample dilution takes place, the RODAC method is suitable for sampling precleaned or sanitized surfaces. If the surface is heavily contaminated, incubated RODAC plates will be crowded with colonies and results will be difficult to interpret.

Air Sampling Microorganisms may become airborne due to activities such as water spraying, dry ingredients handling, and vigorous air movements. Air-suspended dust particles can carry microorganisms. Mold and bacterial spores are common contaminants of air since they survive dryness and other detrimental environmental factors. The microbiological quality of air in a processing facility impacts the quality and safety of perishable food processed in this facility. Improper filtration of air entering this facility or recycling air from the raw product area into the finished product area can result in food contamination. Air quality in the packaging area is particularly important for the control of postprocessing contamination. Therefore, determining the microbial load in air is an important measurement.

Sedimentation is a simple method to measure air quality. It involves exposing agar media plates to air by leaving these plates uncovered in the location to be sampled. Air contaminants will sediment by the force of gravity during the exposure time (e.g., 15 min). The plates are incubated and the colony count may be considered proportional to air contamination level.

Air in a particular environment may be impacted onto the surface of agar media plates using mechanical means. Jets of air are directed over the media plates so that air load collides and sticks to the agar surface. After receiving a measured air sample, the agar plates are incubated and colonies are counted. Air stream also may be filtered through a microfilter. Microorganisms are released from the filter using a suitable diluent and the microbial load is counted.

OBJECTIVES

1. Practice sampling the food processing environment.
2. Determine the total, yeast and mold, and *Pseudomonas* counts in selected food processing environments.
3. Learn about the prevalence of spoilage microbiota and pathogenic bacteria in the food processing environment.

PROCEDURE OVERVIEW

In this exercise, a swab procedure for examination of surfaces will be followed. Two swab samples will be collected: One will be analyzed for total plate count and counts of selected spoilage microorganisms (i.e., yeasts, molds, and *Pseudomonas* spp.) and the other will be tested in a subsequent exercise (Chapter 8) for the presence of a selected pathogen (i.e., *L. monocytogenes*).

Students will take the supplies needed to sample food processing equipment and environment. These supplies include diluent tubes and swabs. Sampling sites may include food processing factories or pilot facilities, chillers, coolers, freezers, food or ingredient storage or holding tanks, packaging machines, meat slicers in deli shops, floors, walls, and drains. A piece of equipment or the interior of a pipe may also be sampled. The choice of sampling sites will be assigned by the instructors. Swab samples are taken according to the protocol indicated later. Test tube samples are promptly refrigerated and analyzed during the subsequent laboratory period.

During the second laboratory period, the environmental samples are analyzed for total plate count, yeast and mold count, and *Pseudomonas* count, and plates are incubated at the appropriate temperature for the tested microbial group. In the third laboratory period, environmental microbiota will be examined and colonies in the incubated plates are counted. The overall scheme of sampling and analysis is shown in Fig. 6.1.

MEDIA

Brain–Heart Infusion (BHI) Broth

This is a rich medium that is frequently used for cultivation of fastidious bacteria and fungi. Because this medium is made from infusions of calves' brain and cattle's heart, BHI agar contains a variety of micronutrients and vitamins. The medium is buffered with disodium phosphate and the final pH of the broth is 7.4. A double-strength (2×) BHI broth will be mixed with an equal volume of an environmental sample so that a normal-strength broth will result. This medium–environmental sample mixture will be cold enriched for detection of *L. monocytogenes*.

Plate Count Agar with and without Antibiotic

These media were explained in previous laboratory exercises (see Chapters 2 and 3).

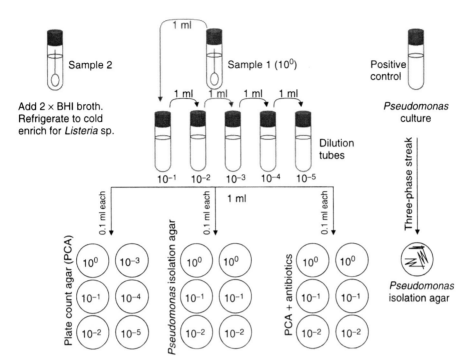

Fig. 6.1. Outline of procedure to determine total, *Pseudomonas*, and yeast and mold counts in environmental samples.

Pseudomonas Isolation Agar (PIA)

This is a selective medium for isolation of pseudomonads. The medium is somewhat minimal, using a carbon source (glycerol) not commonly utilized by bacteria. Few species other than *Pseudomonas* will grow on this medium due to inclusion of Irgasan, a broad-spectrum antimicrobial agent. In addition, the inclusion of magnesium chloride and potassium sulfate in the medium is specifically done to enhance the production of pyocyanin. Pyocyanin is a fluorescent pigment produced by *Pseudomonas aeruginosa*. This pigment is blue to blue-green when viewed using ultraviolet light. The pigment usually diffuses into the medium. Fluorescein is a yellow to yellow-green fluorescent pigment that may be produced by many *Pseudomonas* spp. Pigment production and color may be used to differentiate pseudomonads.

Sterile 0.85% Saline

This is an osmotically balanced salt medium which does not contain any nutrients. It is used here to hold the microbial cells temporarily in stasis so that the microbial population does not change in size between the sampling and plating events. Sterile saline may be used to rinse the swab, but a sterile buffered rinse solution is preferred. Alternatively, a liquid medium designed to neutralize not only acids and bases but also oxidants and phenolic compounds is the most suitable for rinsing and transporting environmental samples.

ORGANIZATION

Each student will take sampling materials (two test tubes, each containing 5 ml sterile 0.85% saline or buffered rinse solution, and two sterile individually packaged swabs) and obtain two environmental samples from the same sampling site. Both samples are brought to the laboratory and refrigerated until analyzed. In the subsequent period of the laboratory exercise, one sample will be directly tested for total count and for counts of *Pseudomonas* spp. and yeasts and molds. The second sample will be cold enriched and tested for *Listeria* sp. in a later exercise (Chapter 8).

REFERENCES

Charlton, B. R., H. Kinde, and L. H. Jensen. 1990. Environmental Survey for *Listeria* Species in California Milk Processing Plants. *J. Food Prot.* 53:198–201.

Harrigan, W. F. 1998. *Laboratory Methods in Food Microbiology*, 3rd ed. Academic, New York, NY.

Evancho, G. M., W. H. Sveum, L. J. Moberg, and J. F. Frank. 2001. Microbiological Monitoring of the Food Processing Environment. In F. P. Downes and K. Ito (Eds.), *Compendium of Methods for the Microbiological Examination of Foods*, 4th ed. (pp. 25–35). American Public Health Association, Washington, DC.

Period 1 Sampling Food Processing Environment

MATERIALS AND EQUIPMENT

Per Student

- Two screw-capped test tubes, each containing 5 ml sterile 0.85% saline
- Two sterile cotton swabs, individually wrapped

PROCEDURE

The environmental sample should be taken at the shortest possible period before the start of the experiment. This minimizes the effect of storage on the microbial population in the sample.

1. Determine a suitable sampling site. It is preferred that both swab samples come from the same general site (e.g., a refrigerator's fruit bin and meat bin). Any food-related equipment or contact surface may be sampled. Potential sampling sites include home kitchens and university food processing pilot plants. Refrigerators (door interior, door gasket, inner surfaces, and shelves) are good choices of sampling sites. Other possible sites would be cutting boards and kitchen counter tops. Do not take environmental samples from places such as supermarkets or restaurants without prior consent and arrangements with their management.
2. Sampling procedures for measurable (flat or slightly curved) surfaces such as doors, bins and shelves, and equipment parts (not readily measurable, such as corners, nozzles, and valves) are slightly different.
   i. Measurable surfaces procedure
      a. At the time of sampling, moisten a swab head in the broth and gently press out the excess solution against the interior wall of the test tube.
      b. Rub the swab head slowly and thoroughly over an area of about $50\,cm^2$ of the sampled surface (e.g., 2 cm wide and 25 cm long).
      c. Rinse the swab head in the broth and press out excess solution.
      d. Repeat the sampling for an additional 50-cm^2 area (on the same general surface) using the same swab. Don't go back over a previously swabbed area.
      e. Break off (or cut) the swab so that only the swab head remains in the tube and the tube can be screwed closed. This ensures that the organisms trapped on the swab itself become part of the enrichment.
   ii. Equipment parts
      a. At the time of sampling, moisten a swab head in the solution and gently press out the excess solution against the interior wall of the test tube.
      b. Swab one-fifth of the part or site being sampled.
      c. Rinse the swab head in the broth and press out excess solution.
      d. Repeat the sampling on the remaining sections using the same swab (swab one-fifth of the area at a time) until the part is completely swabbed.

e. Break off (or cut) the swab so that only the swab head remains in the tube and the tube can be screwed closed. This ensures that the organisms trapped on the swab itself become part of the enrichment.
3. Use one swab and one test tube for each of the two areas sampled.
4. Store samples in the refrigerator. Samples may be brought to the laboratory and refrigerated prior to use.

Period 2 Analysis of Environmental Sample

Environmental samples will be analyzed for total, *Pseudomonas* spp., and yeast and mold counts.

MATERIALS AND EQUIPMENT

Per Student

- Refrigerated environmental sample
- Culture of *Pseudomonas* sp. (positive control)
- Five 9-ml peptone water tubes (diluent)
- Six plate count agar plates (one plate per dilution)
- Seven *Pseudomonas* isolation agar plates (six for plating three dilutions in duplicates and one for the positive control)
- Six plates of plate count agar–antibiotics for yeast and mold counts
- One 5-ml tube of 2× brain heart infusion broth

Class Shared

- Refrigerator, set at 4–6°C
- Incubator, set at 30°C
- Incubator, set at 35°C
- Colony counter

PROCEDURE

Dilutions

1. One of the two swab samples will be used for this analysis (the other is used for the exercise on *Listeria* as described later).
2. Mix the sample by rolling in between hands.
3. Label diluent tubes and prepare 10^{-1}, 10^{-2}, 10^{-3}, 10^{-4}, and 10^{-5} dilutions; follow the guidelines for preparing dilutions as detailed in Chapter 1.

Plating

Total Count in Environmental Sample

1. Spread 0.1 ml of the undiluted sample and 0.1 ml from each dilution tube onto PCA plates (total of six plates, duplicate plating is not done to simplify the procedure).
2. Incubate the plates at 35°C for 24 hr. Alternatively, incubation may be done at 30°C for 48 hr; this allows growth of both mesophilic and psychrotrophic microorganisms.

Detection and Enumeration of **Pseudomonas** *spp.*

1. Spread 0.1 ml of the undiluted sample and 0.1 ml from the 10^{-1} and 10^{-2} dilution tubes onto duplicate plates of the PIA medium.

2. Three-phase streak (for isolation) the positive-control culture onto one PIA plate. Isolation enhances visualization of fluorescent pigments. Each student will do a three-phase streak of the control.
3. Incubate the plates at 30°C for 48 hr.

Yeast and Mold Count
1. Spread 0.1 ml of the undiluted sample and 0.1 ml of the 10^{-1} and 10^{-2} dilution tubes onto the PCA-antibiotic duplicate plates (six plates, total).
2. Incubate the plates at 30°C for 5 days.

Preparation of Cold Enrichment Sample
1. Bring the volume of the remaining environmental sample up to 9 mL using 2× BHI broth. If 4 ml saline–sample mixture is assumed remaining in the tube, add 4 ml 2× BHI broth to the sample to get normal-strength BHI broth.
2. Refrigerate the diluted sample at 4°C for cold enrichment and testing for *Listeria* in a subsequent exercise (Chapter 8).

Period 3 Enumeration of Environmental Microbiota

MATERIALS AND EQUIPMENT

Per Student

- Plates from Period 2
- Goggles for eye protection against UV radiation
- Latex gloves

Class Shared

- UV lamp

PROCEDURE

Total Aerobic Plate Count

1. Inspect the PCA plates. Observe various colonies morphologically. Determine the dominant colony type, unusual colony types, and whether there is a variation in the types of colonies between plates at different dilutions.
2. Count the colonies and report the results in Table 6.1.
3. Calculate total CFU/100 cm^2 surface area or CFU per site or equipment part; record the results in Table 6.2 and show the calculations.

Pseudomonas spp.

1. Inspect the PIA plates visually and observe the similarity or variation in the morphology of colonies.
2. Count *Pseudomonas* colonies. Assume that all colonies growing on the plate are indeed *Pseudomonas* spp. Record the colony counts in Table 6.1.
3. Calculate the results as *Pseudomonas* CFU/100 cm^2 surface area or *Pseudomonas* CFU per site or equipment part; record the results in Table 6.2 and show the calculations.
4. Take the plates into a darkened room to observe them using UV light. This will cause the pigments fluorescein and pyocyanin to glow with characteristic color. Do this under the supervision of a laboratory instructor. Eye protection (UV-absorbing goggles or eyeglasses) and latex gloves are required. Describe the appearance and relative abundance of those colonies that fluoresced.

Yeast and Mold Count

1. Inspect the PCA–antibiotic agar plates visually and observe the similarity or variation in the morphology of colonies.
2. Count yeast and mold colonies. Assume all colonies growing on the plate are yeasts and molds. An example of a yeast and mold plate is shown in Fig. 6.2.

TABLE 6.1. Counts of Colonies from Environmental Sample (List the Source) **When Different Dilutions Were Plated on Plate Count Agar, *Pseudomonas* Isolation Agar, and Plate Count Agar with Antibiotics and After Incubation of Plates Under Conditions Appropriate for Test**

Dilution Factor	Number of Colonies				
	PCA[a]	PIA[b]		PCA–Antibiotics[c]	
		Plate 1	Plate 2	Plate 1	Plate 2
10^0					
10^{-1}					
10^{-2}					
10^{-3}		X[d]	X	X	X
10^{-4}		X	X	X	X
10^{-5}		X	X	X	X

[a] Plate count agar; plates were incubated at 35°C for 48 h.
[b] *Pseudomonas* isolation agar; plates were incubated at 30°C for 48 h.
[c] Plates were incubated at 30°C for 5 days.
[d] Dilution not plated.

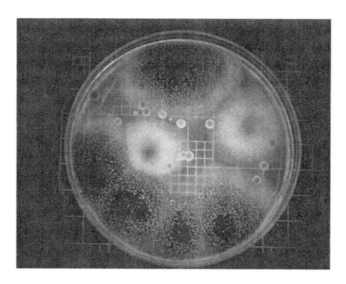

Fig. 6.2. Yeast and mold biota resulting from a swab sample of a vegetable drawer of a home refrigerator. The sample was plated on plate count agar with antibiotics and the plate was incubated at 30°C for 5 days.

TABLE 6.2. Counts of Total, *Pseudomonas*, and Yeast and Mold Populations in Environmental Sample (List Sample Source)

Site Sampled	Site Description	PCA^a	PIA^b	PCA–Antibiotics
		Total plate count = CFU^c/100 cm^2	*Pseudomonas* count = CFU/100 cm^2	Yeast and molds count = CFU/100 cm^2

^{a} Plate count agar.
^{b} *Pseudomonas* isolation agar.
^{c} Colony-forming units.

3. Record the colony counts in Table 6.1.
4. Calculate the results as yeast and mold CFU/100 cm^2 surface area or yeast and mold CFU per site or equipment part; record the results in Table 6.2 and show the calculations.

PROBLEMS

1. What is the significance of microbiological sampling of the environment to the food processor?

2. Why would environmental samples be of more use in a processing plant or a food service area than in a grocery store?

3. What is the source of the environmental sample analyzed in this laboratory exercise? On a scale of 1 to 10 (1 is clean and 10 is dirty), describe the sampled location. For what is this area used? Why was this area chosen for sampling?

4. What was the purpose of refrigerating the environmental samples immediately after sampling before coming to the class? What are the consequences of omitting this step?

5. Would a plating medium other than PCA be better if sampling a milk filler, cheese press, sauerkraut vat, or bread-making equipment? Explain.

6. Compare and contrast the three media (PCA, PCA–antibiotics, and PIA) used, especially with regard to type of organisms likely to grow on each medium, with relevance to the site sampled.

7. Based on results gathered in this laboratory exercise, comment on the microbiological cleanliness of the area sampled. Would raw foods prepared on or with this surface or stored on this surface be safe to eat? What clean-up process would be recommended for this site? Remember that the limited data gathered in this laboratory exercise may not be sufficient to judge the overall safety of the site/part.

8. Explain how environmental data such as what were collected in this laboratory exercise might be used to develop a HAACP plan.

PART III

FOOD-TRANSMITTED PATHOGENS

In this part of the manual, methods to analyze food for selected pathogenic bacteria will be presented. These pathogens include *Staphylococcus aureus*, a gram-positive coccus; *Listeria monocytogenes*, a gram-positive rod; and *Salmonella* spp. and *Escherichia coli* O157:H7, gram-negative rods. Basic techniques learned in Chapter 1 and other practical experiences gained in the subsequent exercises are essential for proper execution of experiments in this part of the manual. Unlike the previous exercises, however, isolation and identification rather than enumeration are emphasized. It is important that students exercise utmost care when working with pathogens. The safety guidelines in the introductory chapter of this manual should be reviewed and understood before running these experiments.

FOODBORNE MICROBIAL DISEASES

Food-transmitted diseases, which are also referred to as foodborne diseases, are any illnesses that result from ingesting food. Health hazards associated with food consumption are caused by physical (e.g., sharp objects), chemical (e.g., heavy metal ions), or microbiological agents. Microbial etiological agents of foodborne diseases (i.e., pathogens) only are addressed in the manual. Food-transmitted diseases of microbial origin are broadly classified into infections and intoxications based on the mode of action of the etiological agents (Fig. III.1). If ingestion of living cells is required to cause the foodborne disease, it is described as infection. Intoxication results from ingestion of toxins produced by microorganisms in food. Ingestion of living cells is not required to cause microbial intoxication; instead, presence of the microbial toxin in ingested food causes the disease. Foodborne infections include diseases caused by invasive or noninvasive means. Noninvasive infections (or toxico-infections) are caused by microorganisms that thrive in the intestine and produce toxins; their close proximity to the intestinal wall enhances their ability to cause the

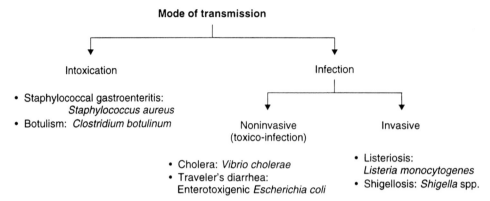

Fig. III.1. Mode of transmission of bacterial foodborne diseases and examples representing each category.

disease. These microorganisms do not need to invade the intestinal wall or body tissues to cause the disease. Invasive infections, however, result from pathogen invasion of the intestinal wall; the pathogen may also reach other body tissues.

Understanding the events that precede transmission of microbial diseases by food may help develop proper control measures. Prerequisites of microbial diseases transmission by food include

1. Presence of the pathogenic microorganism in food.
2. Growth or survival of the microorganism in food.
3. Ingestion of food by susceptible individuals.

In case of infections, *mere presence* and survival of the pathogens in food may be sufficient to cause a disease when this food is consumed. The infectious dose may vary with the pathogen, food, and susceptibility of the individual consuming the food. On the contrary, contamination of food with a small population of a toxigenic microorganism may not be sufficient to cause health hazards if the microorganism is unable to grow in food. Therefore, for intoxication to occur, *presence and growth* of the pathogen in food are prerequisites for disease transmission. Consumers who are most susceptible to foodborne diseases are the very old, the very young (e.g., newly born), the pregnant, and the immunocompromised.

DETECTION OF PATHOGENS IN FOOD

If pathogens are present in food, only a small population (usually $<10^3$ CFU/ml or CFU/g) is expected. Direct plating of such a food reveals little or no information about the pathogen in question since the population of the pathogen in food is most likely below the detection level of the counting techniques. Therefore, most microbiological analyses for foodborne pathogens are designed only to determine their presence or absence in food. These analyses, therefore, are called detection methods. Pathogens that cause infections (e.g., *L. monocytogenes* and *E. coli* O157:H7) should not be present in ready-to-eat food in any detectable levels. Presence of these pathogens in raw food before processing also is not desirable, since contamination of the processing facility with the pathogen may lead to contamination of the finished product.

Presence of a small population of intoxication-causing pathogens in raw food before processing may be tolerated provided that no toxin has already been produced in food and that there is a processing step that eliminates or prevents the growth of the pathogen in the finished product. A small number of *Clostridium botulinum* spores, for example, may be present on raw beans, but the properly retorted canned product should not cause a botulism hazard to consumers. In this case, testing for *C. botulinum* in the raw or finished product may not be justifiable; however, if these are concerns about product safety, detection of the toxin in the cans may be warranted. Some manufacturers of minimally processed food, however, may test raw or processed food for toxigenic microorganisms (e.g., *S. aureus*). A high population of *S. aureus* or *Bacillus cereus* in raw or processed food is alarming to food producers, and thus determining the count of these pathogens in food is carried out occasionally. Additionally, some of the microbial toxins are relatively heat stable and thus may survive processing if produced in the raw product.

Detection Methods

Detection of pathogens in food requires the execution of multistep microbiological analytical *methods* or *protocols*. Each of these methods includes one or more *technique*. Methods for detection of pathogens in food include cultural, biochemical, immunological, or genetic techniques (Fig. III.2) or combinations of these techniques. Detection of *Salmonella* spp. using the Food and Drug Administration (FDA) protocol, for example, involves enrichment and isolation steps (culture techniques), identification (a biochemical characterization technique), and serotyping (an immunological technique). The protocol just described is considered a conventional detection method, and newer rapid alternatives have been developed (Fig. III.3). An alternative rapid detection method for *Salmonella* spp. in food also includes an enrichment procedure, followed directly by an immunological or a genetic detection technique, with no isolation step. Rapid detection methods allow the analyst to rapidly screen negative samples, and this decreases analysis time considerably. Samples that are positive by these methods, however, need to be confirmed by the conventional cultural and biochemical techniques.

One of the goals of the detection method is to identify the pathogen of interest. *Identification* normally refers to naming the microorganism's genus and preferably the species as well. Classification of a microorganism at the subspecies level is known as *typing*. Cultural, biochemical, immunological, or genetic techniques similar to those used in identification are also employed in typing. Additionally, sensitivity of bacteria to different phages is the basis for useful typing methods. If *L. monocytogenes*, for example, was isolated from a food, it may be necessary to "type" the isolate using a ribotyping method. This typing may enable investigators to track a disease and verify the epidemiological link between a suspected food and a disease outbreak.

In all exercises presented in this part of the manual, students will practice the conventional methods with emphasis on the culture techniques for isolation and identification of the targeted pathogen. Advanced immunological and genetic techniques will be included in some of the exercises. The principles of these methods

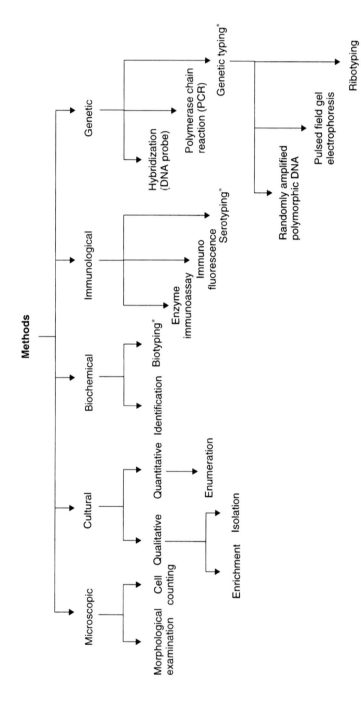

Fig. III.2. Methods commonly used in the microbiological analysis of food. Asterisks refer to typing methods, which characterize microorganisms at the subspecies level.

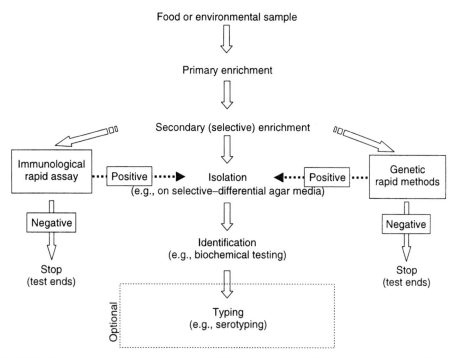

Fig. III.3. Conventional and alternative rapid procedures for detection of pathogens in food (generalized).

are described in this introduction, but details about specific applications are included in the subsequent chapters. When market food is analyzed, it is unlikely that pathogens are detected easily in these products. Professional microbiological analysts commonly find more negative than positive samples, therefore, analyzing these negative samples helps students understand the reality about testing for pathogens in the food industry.

In addition to food samples, students will also run parallel analysis of positive and negative controls. A *positive control* is a microorganism similar to that targeted by the analysis. Therefore, if food is analyzed for *S. aureus*, the positive control would be a culture of *S. aureus* with the typical properties of this microorganism, that is, lecithinase, catalase, and thermonuclease positive and capable of fermenting mannitol anaerobically. Because of safety concerns, a nonvirulent strain may be used as a positive control provided that the strain has similar properties and produces most of the typical reactions of the microorganism of interest. On the contrary, a *negative control* is a microorganism that does not produce typical reactions of the one targeted by the analysis, but it still grows on most media used in the analysis. Analyzing the positive and negative controls allows the analyst to verify the correctness of the procedure and to test the quality of media and reagents.

Method Sensitivity and Specificity

Sensitivity is the proportion of positive test results obtained when the method is applied to samples known to carry the microorganism targeted by the analysis. To illustrate the concept, suppose that 100 food samples have been inoculated with a pathogen and analyzed by a method designed to detect this pathogen. If the method detects the pathogen in 97 of these samples, method sensitivity is 97%. This leaves three samples that tested negative while in fact these are supposed to be positive. Therefore, the method produces a false-negative rate of 3%. Method sensitivity also determines the minimum detectable concentration of the targeted microorganism in the contaminated sample. A method with high sensitivity detects pathogens present in smaller numbers in the sample compared to a method with low sensitivity.

Specificity defines the method's ability to distinguish a targeted microorganism in the tested sample from other microorganisms in the sample. For example, if 100 food samples containing natural microbiota were free from a pathogen of concern but the analytical method produced positive results in 5 samples, method specificity is 95%. The method also is described as producing a false-positive rate of 5%.

SELECTED METHODS

Culture-Based Methods

Culture techniques are widely used in the microbiological analysis of food. Conventional procedures, including culture and biochemical methods, constitute the "gold standard" for microbiological analysis of food. Validation of novel detection methods, such as those based on immunology or genetics, depends on the agreement of these new methods with the conventional counterparts.

Culture techniques involve mixing samples to be analyzed with suitable broth or agar media and incubating the mixtures under suitable condition so that microorganisms in these products (e.g., food) grow and produce colonies or reactions. The key feature of these techniques, therefore, is culturing the microorganism or the microbiota in question. When agar media are used, visual observation of incubated plates to determine *colony morphology* should follow. A microscopic examination of cells constituting these colonies may provide valuable *cell morphology* information for confirming the visual observation. When microbial contents of food samples are cultured in broth media, typical reactions of the microorganism or the microbiota in question are usually sought. Part II of this manual includes several examples of typical culture techniques.

Culture techniques are carried out to enrich, enumerate, or isolate investigated microorganisms. *Enrichment* is done when the microorganisms in question cannot be detected by direct plating of food samples. Resuscitation of injured cells in the sample is another reason for applying the enrichment techniques. A multistep enrichment technique usually includes culturing in a nonselective broth medium (preenrichment or primary enrichment) followed by subculturing in a selective broth medium (selective or secondary enrichment). Conditions conducive to rapid growth of sample microbiota may be avoided in the preenrichment step. Therefore, during preenrichment, it is common to avoid using extremely rich media or incubation conditions that favor rapid multiplication. During the preenrichment step, both the microorganism of interest and the food microbiota will increase in number. Selective or secondary enrichment involves transferring a portion of the preenrichment into a selective or selective–differential broth medium. During this stage, only the microorganism of interest will grow while growth of other microorganisms is suppressed.

Enumeration of a microorganism or a group of microorganisms in the food sample is one of the most common reasons for using culture techniques. In this case, nonselective or selective media are used and colonies are inspected and counted. Alternatively, selective–differential liquid media are used to determine the size of the microbial population in the sample using the MPN technique. Instrumentation-assisted culture techniques have been developed to speed the enumeration process or to produce objective and consistent counts. The spiral plating method is an example of such innovations. Determining cell density by measuring culture turbidity or absorbance is a simple technique, but results cannot be reliably converted to counts (i.e., CFU/ml).

Isolation of a microorganism from food is possible when the sample, with or without enrichment, is plated on a suitable selective or selective–differential medium. Culturing on these media is currently the most reliable method to isolate a targeted microorganism. Use of microbiological media in food analysis was addressed in an earlier chapter, and several selective and selective–differential media will be used in exercises presented in this part of the manual.

Biochemical Methods

Biochemical methods have been used for decades as identification techniques that succeed in isolating microorganisms on culture media. These methods measure specific microbial metabolic activities and thus help identify targeted microorganisms. For conventional biochemical testing, cultures in broth or well-isolated colonies from agar media plates (i.e., isolates) are mixed with reagents in suitable media and mixtures are incubated. Production of specific metabolites is detected when they react with media reagents. Biochemical methods can be as simple as transferring a portion of a colony to a slide, adding a drop of hydrogen peroxide solution, and observing gas bubble formation. Formation of gas bubbles, in this case, indicates that the isolate is catalase positive. Biochemical identification methods will be used in subsequent chapters of this manual. For example, colonies isolated from food and believed to be *Salmonella* spp. are inoculated into triple sugar iron (TSI) agar for biochemical identification. If hydrogen sulfide (H_2S) is produced during incubation, it reacts with ferrous sulfate ($FeSO_4$) in TSI, resulting in a black precipitate. Production of this metabolite (i.e., H_2S) under the test conditions is indicative of *Salmonella* spp.

In recent decades, great progress has been achieved in developing rapid and elaborate biochemical identification techniques. One of these innovations (i.e., the API diagnostic system) uses a strip of wells containing small portions of dehydrated media. Addition of a suspension of the microorganism in question hydrates the media in the wells. The strip is incubated and different reactions in wells are recorded. Results of these biochemical reactions are compared to charts or analyzed by computer programs to determine the identity of the isolate.

Immunological Methods

Immunological methods in microbiology rely on the fact that animals produce specific *antibodies* when injected with microbial cells or some components of these cells (i.e., *antigens*). Blood serum from these animals can be prepared for use in the laboratory to react with any isolate with great similarity to the original microorganism that the animal was exposed to earlier. Antibodies, with a great degree of specificity to particular antigens, now can be produced from tissue cultures for diagnosing microorganisms such as those found in food. Additionally, improved assay techniques and methods to detect the antigen–antibody interactions have been developed.

Serotyping is an immunological method used to classify microorganisms at the subspecies level into serotypes. Serotyping is particularly important in the identification of *Salmonella* in food since this genus includes hundreds of serotypes that are difficult to differentiate by other techniques.

Enzyme immunoassay, or more specifically *enzyme-linked immunosorbent assay (ELISA)*, is a popular immunological method for detection of pathogens. While indirect ELISA is used to detect antibodies produced against the pathogen of concern (e.g., in blood samples), direct ELISA detects the antigens of the pathogen. ELISA can be used not only to detect but also to quantify a targeted antigen. In one of the direct ELISA formats (Fig. III.4), plastic wells are coated with the specific antibody against the targeted microorganism. When the sample containing the target antigen is added to the wells, the target will bind to the antibody on the well, creating an immune complex that remains attached to the well's plastic surface. An additional enzyme-labeled (conjugated) secondary antibody is added to bind to the immune complex. Horseradish peroxidase or alkaline phosphatase is commonly used to label the secondary antibody. Presence of the enzyme in the well, after washing, indicates the presence of the target. Presence of the enzyme is detected by addition of an appropriate chromogenic substrate.

Genetic (or Nucleic Acid–Based) Methods

Nucleic Acid Hybridization The genetic information of a bacterial cell is contained in a chromosome and sometimes plasmids, which are made of deoxyribonucleic acid (DNA). The bacterial cell contains a single chromosome whereas the fungal cell contains multiple chromosomes. DNA includes four nucleotide bases: adenine, thymine, cytosine, and guanine; the number and sequence of these bases in the chromosome define the characteristics of the microorganism. Sequences in the DNA molecule that are unique to a particular microorganism may be used in its identification. Since adenine pairs with thymine and cytosine pairs with guanine, it is possible to develop a sequence of nucleotides that hybridizes with a complementary unique segment of a bacterial chromosome. When this unique sequence (e.g., 20–200 nucleotides) is labeled for easy identification, it is called a *DNA probe*. Radioactive isotopes were used to label DNA probes, but enzymes are now more commonly used. If a DNA probe hybridizes with DNA from an unknown isolate, it is possible to detect this event and thus identify the microorganism representing this isolate. Probes that target ribosomal ribonucleic acid (rRNA) are also used in microbial identification. For detection of a microorganism by nucleic acid hybridization methods, 10^5–10^6 copies of target DNA are required. Cells contain an abundance of rRNA, and thus identification methods using rRNA probes are relatively sensitive and efficient. Detection procedures using nucleic acid hybridization methods, in general, should include enrichment steps to increase the population of the microor-

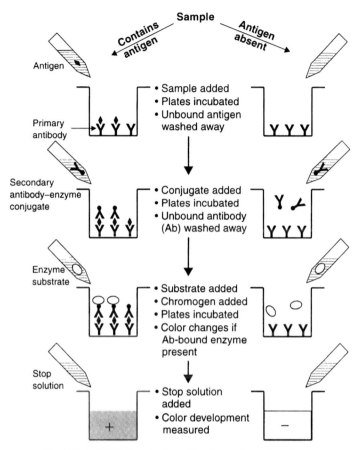

Fig. III.4. ELISA technique for detection of antigens.

ganism of interest and the copies of targeted DNA or RNA. In Chapter 9 of this manual, a hybridization technique will be used to detect *Salmonella* in food.

Polymerase Chain Reaction (PCR) Technique The PCR technique is used to replicate a predefined DNA sequence from the target microorganism (Fig. III.5). Compared to a hybridization technique, a smaller number of copies of target DNA is needed; nevertheless, an enrichment step is needed in the detection procedure. In this technique, DNA from the target microorganism is mixed with a heat-resistant DNA polymerase (e.g., Taq polymerase), nucleotides, and a primer. The primer is a pair of oligonucleotides that are designed to flank a unique sequence on the target DNA. The mixture is heated and cooled in timed cycles, commonly in a thermocycler. When the mixture is heated (~94°C), DNA denatures into single strands. Cooling this mixture allows the primer to recognize and bind to the target DNA (i.e., anneal). Temperature of annealing varies with the primer structure. For optimum functionality of the polymerase, the temperature of the mixture is raised, generally to 68–72°C. At this time, DNA polymerase extends the primer using the free nucleotides; thus two copies of the target DNA are created. Repeated heating

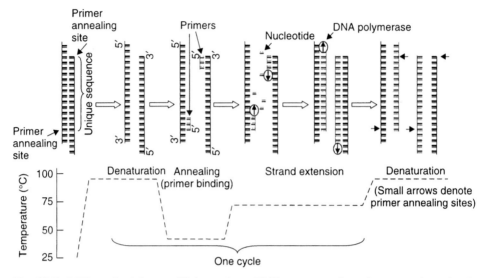

Fig. III.5. PCR method for amplifying unique DNA sequence for subsequent detection by gel electrophoresis.

and cooling creates millions of copies of the target DNA in a short period of time. If the target DNA is absent, the primer cannot anneal (bind) and thus no amplification occurs.

The DNA components of the PCR-amplified sample are separated on agarose gel by electrophoresis. A rectangular slab of agarose gel is prepared with wells on one end of the rectangle. The samples, containing the PCR products, are dispensed (loaded) in the gel wells. A tracking dye and a mass marker (DNA ladder) are also loaded in separate wells. Electricity from a power supply is turned on to allow separation of DNA fragments on the basis of mass and charge. Large fragments move slower than the smaller ones. After the run is complete, the gel is stained with ethidium bromide, a DNA-binding fluorescent dye. Alternatively, ethidium bromide may be mixed into the gel prior to gel loading. The ethidium bromide–DNA complex fluoresces when exposed to UV radiation. A picture is taken of the fluorescent bands. Size of separated bands, in base pairs (bp), is determined relative to the mass marker. Presence of a band with the correct molecular weight indicates the presence of the pathogen in the sample. A PCR-based detection procedure will be tested during the exercise to detect *L. monocytogenes* in food (Chapter 8).

REFERENCES

Fung, D. Y. C. 2002. Rapid Methods and Automation in Microbiology. *Comprehensive Rev. Food Sci. Food Safety* 1:3–22.

Entis, P., D. Y. C. Fung, M. W. Griffiths, L. McIntyre, S. Russell, A. N. Sharpe, and M. L. Tortorello. 2001. Rapid Methods for Detection, Identification, and Enumeration. In F. P. Downes and K. Ito (Eds.), *Compendium of Methods for the Microbiological Examination of Foods*, 4th ed. (pp. 89–126). American Public Health Association, Washington, DC.

CHAPTER 7

Staphylococcus aureus
ISOLATION AND IDENTIFICATION;
USING CULTURE TECHNIQUE;
HANDLING PATHOGENS

INTRODUCTION

Properties

Staphylococcus aureus is a gram-positive coccus with a cell diameter that ranges from 0.5 to 1.5 µm. When seen under the microscope, cells appear in clusters. The bacterium is catalase and coagulase positive and produces yellow-pigmented colonies (Table 7.1). The microorganism is hemolytic, produces a thermonuclease, and ferments mannitol anaerobically. Ability of the bacterium to survive for an extended period in a dry state is an important property that concerns the food industry. It also can grow in food and media with water activity (a_w) as low as 0.86. This pathogen produces an enterotoxin, known as *staphylococcal enterotoxin*, that causes *staphylococcal gastroenteritis*, a food-transmitted disease. The toxin may be produced by other *Staphylococcus* spp., but the disease is almost always associated with *S. aureus*. Some strains of *S. aureus*, however, may not produce the toxin. The toxin is a protein with several variants or antigenic types.

The Disease

Staphylococcus aureus causes several human illnesses, including a food-transmitted disease. Staphylococcal gastroenteritis, which was commonly known as staphylococcal food poisoning, probably has one of the highest incidence rates among food-transmitted diseases. The disease is an intoxication that results from consumption of food where *S. aureus* grew and produced the toxin. Population of *S. aureus* in food usually grows to at least 10^6 CFU/ml or CFU/g before the toxin is produced to levels that can cause the disease. Ingestion of less than 1 µg of the toxin may cause

---

Food Microbiology By Ahmed E. Yousef and Carolyn Carlstrom
ISBN 0-471-39105-0 Copyright © 2003 by John Wiley & Sons, Inc.

TABLE 7.1. Characteristics of Selected *Staphylococcus* Species

Property	S. aureus	S. epidermidis	S. hyicus	S. intermedius
Coagulase	+	−	w	+
Heat-stable (thermo) nuclease	+	−	+	+
Yellow pigment	+	−	−	−
Hemolysis	+	v	−	+
Mannitol fermentation anaerobically	+	−	−	−

Key: (+) positive, (−) negative, (v) variable, (w) weak reaction.

the disease. The symptoms of the disease appear shortly after ingestion of contaminated food; onset time is only 4–6 hr. The most common symptoms are vomiting and abdominal pain. Nausea and diarrhea are also common, while fever is absent. The disease is self-limiting, with symptoms subsiding in a few hours, and it rarely causes death. Children and the elderly are the most susceptible individuals.

Incidence in Food

Foods commonly associated with staphylococcal food intoxication include cream-filled bakery products, dairy products, salads, and some meat products. The human nose and throat may contain *S. aureus*, and thus improper handling of processed food may lead to its contamination with the pathogen from the workers. Presence of high counts of *S. aureus* in food ($>10^5$ CFU/ml or CFU/g) poses a health hazard to consumers. Presence of low counts may indicate that the hazard of staphylococcal food intoxication is not imminent. However, if food properties and storage conditions are conducive to the growth of the pathogen, toxin may be produced and the food becomes unsafe to consume. Additionally, low counts of *S. aureus* in food may result from inactivation of a large population of the pathogen by processing means. Most of these processing treatments, however, do not eliminate the toxin. The toxin remains active after heat treatments similar to pasteurization. Heating food at 100°C for 30 min does not completely inactivate the toxin. Pasteurization of milk (e.g., 71.6°C for 15 sec) is sufficient to eliminate the microorganism but not the toxin.

If food is analyzed for *S. aureus*, it is more common to enumerate than just to detect the pathogen. Food implicated in staphylococcal gastroenteritis often contains a large population of *S. aureus*, and thus enumerating the pathogen in such a food may reveal the cause of a disease outbreak. Analysis for the toxin in food may be useful in this case, but variants of the toxin may escape detection. The analysis for *S. aureus* may be done routinely, and the presence of a large count of *S. aureus* may alert the food processor to take corrective measures.

Isolation and Identification

Some of the important features that can be used to detect *S. aureus* include growth in the presence of up to 10% salt (NaCl), 0.5% lithium chloride (LiCl), or 0.05%

potassium tellurite (K_2TeO_3); fermentation of mannitol anaerobically; reduction of tellurite to produce black colonies; hydrolysis of egg yolk (due to the production of lecithinase); and production of coagulase and thermonuclease. Some of these properties are used to design selective–differential media suitable for isolation and enumeration of *S. aureus*. Baird–Parker agar is a selective–differential medium that is commonly used for isolation of *S. aureus*. Coagulase and thermonuclease tests are commonly used to confirm that typical isolates from Baird–Parker agar medium are of *S. aureus*.

OBJECTIVES

1. Determine the count of *S. aureus* in food.
2. Practice isolation technique using selective and differential media.
3. Practice safe procedure for handling pathogens in the laboratory.

PROCEDURE OVERVIEW

An overview of the procedure is shown in Fig. 7.1. During the first period of this exercise, selected foods will be homogenized, additional dilutions are made, and dilutions are plated on Baird–Parker agar medium. To enhance the analyst's ability to detect a small population of *S. aureus* in the analyzed food, a relatively large volume (1 ml, compared with 0.1 ml in traditional spread plating) of each dilution is divided and spread on three Baird–Parker agar plates (Fig. 7.2). After plates are incubated, typical and suspect *S. aureus* colonies will be counted (Period 2). A typical *S. aureus* colony is circular, 2–3 mm in diameter, jet-black to gray-black, and surrounded with an opaque halo and a clear zone (Fig. 7.3). Colonies that are gray or those without halos or clear zones may be considered suspect *S. aureus*. Typical and suspect colonies will be counted separately and these presumptive counts will be confirmed or excluded after additional testing (Periods 2–4). After confirmation, counting rules will be applied, as discussed earlier.

To confirm the identity of the counted colonies as *S. aureus*, typical and suspect isolates from Baird–Parker agar medium are streaked on mannitol salt agar (MSA) and plates are incubated anaerobically (Period 2). Methods to achieve anaerobic conditions were discussed in Chapter 5. Incubated plates are examined for colonies that developed, in spite of the high salt content of this medium, and fermented mannitol anaerobically (Period 3). Additionally, typical and suspect *S. aureus* colonies on Baird–Parker agar medium are streaked on brain–heart infusion (BHI) agar, and colonies on the incubated plates are examined morphologically. Colonies of *S. aureus* typically produce yellow pigmentation on this medium.

Additional confirmation and characterization of *S. aureus* will be carried out in Periods 3 and 4 (Fig. 7.1). Isolates that produced typical reactions on BHI agar and MSA (Period 3) are tested for coagulase and thermonuclease activities. Therefore, typical *S. aureus* isolates are transferred from BHI plates into coagulase plasma ethylenediaminetetraacetic acid (EDTA) tubes or heated and streaked onto toluidine blue DNA agar plates. Coagulase results may be read during the same laboratory period, whereas toluidine blue DNA agar plates are incubated and examined later (Period 4). Isolates that coagulate rabbit plasma within 6 hr of incubation at 35°C

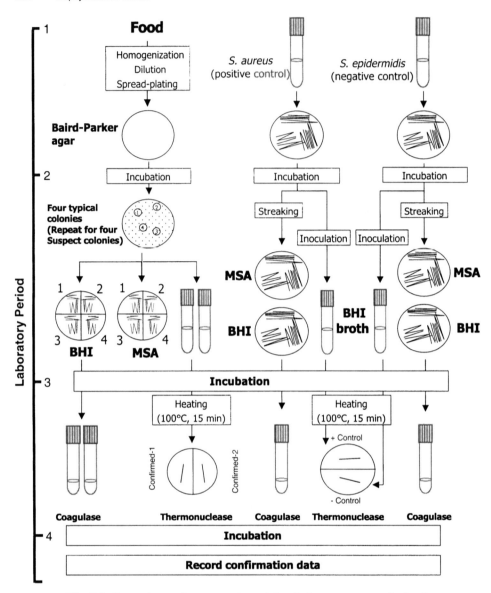

Fig. 7.1. Detection and enumeration of *Staphylococcus aureus* in food.

are considered coagulase positive, and those that produce pink halos around the streak on toluidine blue DNA agar are considered thermonuclease positive.

MEDIA

Baird–Parker Agar

This is a selective–differential medium specifically designed for the detection and enumeration of *Staphylococcus* spp. *Staphylococcus aureus* forms black, shiny,

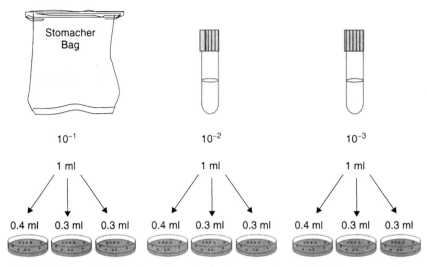

Fig. 7.2. Spread plating large inocula for improved sensitivity in enumeration.

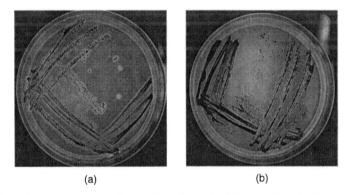

Fig. 7.3. Colonies of *S. aureus* (A) and *S. epidermidis* (B) on Baird–Parker agar medium.

convex colonies surrounded by a clear zone. Other species may produce gray or less shiny black colonies. Lithium chloride and potassium tellurite in this medium select against most non–*Staphylococcus* spp. Differentiation of *Staphylococcus* spp. is brought about by tellurite and egg yolk. *Staphylococcus aureus* reduces the tellurite salt to elemental tellurium, resulting in the formation of black colonies. The microorganism produces lecithinase enzyme that hydrolyzes egg yolk lipids, which results in the clear zone around the *S. aureus* colonies. *Staphylococcus epidermidis* is lecithinase negative.

Brain-Heart Infusion (BHI) Broth or Agar

See Chapter 6 for detailed description. Isolates will be subcultured on this medium to ensure purity of the culture and to produce colonies suitable for coagulase testing.

Coagulase Plasma EDTA

This is a lyophilized rabbit plasma to which EDTA was added as an anticoagulant. Coagulase is an enzyme that coagulates plasma from humans, horses, rabbits, and other animals. It binds with fibrinogen in the plasma, causing it to coagulate. Coagulase activity is tested by mixing the culture into rehydrated plasma. If coagulation (clotting) occurs within 6 hr, this indicates that coagulase is produced by the culture.

Mannitol Salt Agar

This is a selective–differential medium used to isolate *Staphylococcus* spp. *Staphylococcus aureus* grown on MSA will turn the medium a bright yellow color. The salt (7.5% NaCl) in the medium limits the growth of organisms to those that are salt tolerant, such as *Staphylococcus* spp. The combination of phenol red (a pH indicator) and mannitol (a carbohydrate) allows the differentiation of species that can ferment this sugar. If fermentation occurs, acid byproducts result and the pH indicator turns yellow. Organisms unable to ferment the mannitol will not produce significant levels of acid. These organisms may grow on the medium using protein, but the agar will retain the original color or turn pink due to production of basic end products. Use of anaerobic chambers allows differentiation of aerobic versus anaerobic use of mannitol. Organisms such as *S. epidermidis* that are unable to ferment mannitol or use protein anaerobically will not grow under these conditions.

Toluidine Blue DNA Agar

This is a differential medium that is used for detection of deoxyribonuclease of *S. aureus* and other microorganisms. It contains DNA and toluidine. To test for thermonuclease activity, an overnight culture is heated at 100°C for 15 min and streaked (in one straight line) onto the toluidine blue DNA agar. If the culture produces a thermonuclease (i.e., heat-resistant nuclease), the enzyme cleaves DNA and the released nucleotides react with toluidine, which changes from blue to red around the streak.

ORGANIZATION

Each pair of students will test one food. Possible foods include fermented sausages, tuna salad, potato salad, Camembert cheese, or frozen bread dough. Previously described procedures for food sampling, homogenization, and dilution will be followed in this exercise (Fig. 7.1).

Staphylococcus aureus is an infectious and toxigenic bacterium that should be handled with care. Follow the safety guidelines that were reviewed early in this manual. Use disposable gloves when handling the isolates and the pathogenic cultures. Make sure to sanitize the work area after use.

REFERENCES

Jay, J. M. 2000. *Modern Food Microbiology*. Aspen Publishers, Gaithersburg, MD.

Kloos, W. E., and K. H. Schleifer. 1986. *Staphylococcus*. In P. H. Sneath, N. S. Mair, M. E. Sharpe, and J. G. Holt (Eds.), *Bergey's Manual of Systematic Bacteriology*, Vol. 2 (pp. 1013–1035). Williams & Wilkins, Baltimore, MD.

Lancette, G. A., and R. W. Bennett. 2001. *Staphylococcus aureus* and Staphylococcal Enterotoxins. In F. P. Downes and K. Ito (Eds.), *Compendium of Methods for the Microbiological Examination of Foods*, 4th ed. (pp. 387–403). American Public Health Association, Washington, DC.

Period 1 Sampling, Homogenization, and Plating

During this period, the selected food will be homogenized and the 10^{-2} and 10^{-3} dilutions are made. Dilutions are spread plated on Baird–Parker agar medium using larger than normal inoculation volume (Fig. 7.2). Plates will be incubated at 35°C for 48 hr and examined during the subsequent laboratory period.

MATERIALS AND EQUIPMENT

Per Pair of Students

- Positive control: *Staphylococcus aureus* overnight culture in a test tube
- Negative control: *Staphylococcus epidermidis* overnight culture in a test tube
- One 99-ml peptone water
- Two 9-ml peptone water blanks
- Eleven Baird–Parker agar plates

Class Shared

- Scale for weighing food samples (e.g., a top-loading balance with 500 g capacity)
- Stomacher and stomacher bags
- Incubator, set at 35°C

PROCEDURE

Dilutions

1. Prepare the food for sampling to ensure that a representative sample is analyzed. This varies with the food and may involve cutting, partitioning, grinding, or mixing.
2. Weigh 11 g of food directly into a stomacher bag, then add 99 ml peptone water.
3. Stomach for 2 min.
4. Prepare 10^{-2} and 10^{-3} dilution tubes using the peptone water blanks.

Plating

Food Sample

1. Label three plates, containing the Baird–Parker agar medium, for each of the 10^{-1}, 10^{-2}, and 10^{-3} dilutions.
2. Dispense 0.4, 0.3, and 0.3 ml (a total of 1 ml) of the 10^{-1} dilution into three plates (Fig. 7.2).
3. Repeat the previous step using the 10^{-2} and 10^{-3} dilutions.
4. Spread the inocula as indicated previously for the spread plating technique (Chapter 1). Avoid using freshly prepared agar plates; the relatively large inoculum in this exercise will be difficult to spread evenly on such a moist agar.

5. Keep the plates undisturbed for approximately 10 min and make sure the inoculum is absorbed by the agar. Invert plates and incubate at 35°C for 48 hr.

Controls

1. Streak, for isolation, a positive (*S. aureus*) and a negative (*S. epidemidis*) control. Use a separate Baird–Parker agar plate for each organism.
2. Invert the plates and incubate at 35°C for 48 hr.

Period 2 Enumeration and Plating for Confirmation

Typical *S. aureus* colonies are counted; these are circular, 2–3 mm in diameter, jet-black to gray-black, and surrounded with opaque halos and clear zones (Fig. 7.3). Colonies that are gray or those without halos or clear zones are considered suspect *S. aureus*. Typical and suspect colonies will be counted, separately, and these separate counts will be confirmed or excluded after additional testing. After confirmation, counting rules will be applied, as discussed in Chapter 1.

MATERIALS AND EQUIPMENT

Per Pair of Students

- Incubated Baird–Parker plates
- Four BHI agar plates
- Four MSA plates
- Six BHI broth tubes (9 ml each)

Class Shared

- Colony counter
- Anaerobic system: Anaerobic jar, reagent packet, and anaerobic indicator
- Incubator, set at 35°C

PROCEDURE

Counting

1. Inspect the Baird–Parker agar plates from the positive-control culture and make sure that the colonies show the typical *S. aureus* morphology.
2. Inspect the Baird–Parker agar plates from the negative-control culture to become familiar with nontypical morphology.
3. Inspect the Baird–Parker agar plates from the food sample and determine if typical *S. aureus* colonies are present (see the previous morphological description).
4. Divide the colonies on the sample plates into a least three morphological groups based on the degree of similarity to typical *S. aureus* colonies: (a) typical, (b) suspect, and (c) nontypical.
5. Count the colonies with typical and suspect morphology separately. Report these preliminary results in Table 7.2.
6. Calculate presumptive *S. aureus* CFU/g.
    a. Remember that a total of 1.0 ml was plated at each dilution. Therefore, sum the counts for the plates at each dilution. Determine CFU/g using totals that are in the 20–200 range. If more than one total is in this range, average the results. If no results are in this range, use the appropriate rules to estimate the CFU/g.

TABLE 7.2. Count of Typical and Suspect *S. aureus* Colonies on Baird–Parker Agar Plates Obtained by Plating Dilution of the Homogenized Food Sample and Incubating the Inoculated Plates at 35°C for 48 h

Dilution Factor	Count of Typical *S. aureus* Colonies				Count of Suspect *S. aureus* Colonies			
	Plate 1	Plate 2	Plate 3	Total	Plate 1	Plate 2	Plate 3	Total
10^{-1}								
10^{-2}								
10^{-3}								
CFU/g								

(Appropriate footnotes might include a description of suspect colonies).

b. Report results of typical *S. aureus* colonies as typical CFU/g and that of suspect colonies as suspect CFU/g.

7. Save the Baird–Parker agar plates at 4°C for the subsequent laboratory period as colony morphologies may need to be reinspected.

Plating for Confirmation

Colonies with Typical S. aureus Morphology

1. Inspect the Baird–Parker agar plates and mark (with numbers) four isolated colonies with typical *S. aureus* morphology.
2. Label one plate each of BHI agar and MSA and two tubes of BHI broth. Divide each plate area into quarters using a marker. Descriptively label each quadrant as required.
3. Streak each of the four isolates onto both BHI and MSA plates. Use the same colony to inoculate both plates. Use a two-phase streak.
4. Inoculate two BHI broth tubes using two of the four marked colonies (step 1).
5. Invert all plates and incubate all media at 35°C for 24 hr. MSA plates are incubated anaerobically (see Chapter 5 for preparation and use of anaerobic jars).

Colonies with Suspect S. aureus Morphology

1. Inspect the Baird–Parker agar plates and mark (with numbers) four isolated colonies with suspect *S. aureus* morphology. If more than one type of suspect colony were identified, then choose at least one colony from each type. If no suspect types were identified, then skip this section.

2. Label one plate each of BHI agar and MSA and two tubes of BHI broth. Divide the area of each plate into quarters, and label each quadrant.
3. Streak each of the four isolates onto both BHI and MSA plates. Use the same colony to inoculate both plates. Use a two-phase streak.
4. Inoculate two BHI broth tubes using two of the four marked colonies (step 1).
5. Invert the plates and incubate all media at 35°C for 24 hr. The MSA plates are incubated anaerobically.

Positive and Negative Controls

1. Label one plate each of BHI agar and MSA and one BHI broth tube for positive control (*S. aureus*).
2. Label one plate of BHI agar and MSA and one tube of BHI broth for the negative control (*S. epidermidis*). Negative control is done in spite of the fact that it does not grow anaerobically on MSA!
3. Three-phase streak colonies from the Baird–Parker agar control plates onto the BHI agar and MSA plates.
4. Inoculate the BHI tubes with colonies from the Baird–Parker agar control plates.
5. Invert the plates and incubate all media at 35°C for 24 hr. The MSA plates are incubated anaerobically.

Period 3 Confirmation

In this period, typical and suspect *S. aureus* colonies from Baird–Parker agar medium will be confirmed or excluded after observing colony morphology on BHI agar and biochemical reactions on MSA. Additional *S. aureus* properties will be checked using coagulase and thermonuclease tests.

MATERIALS AND EQUIPMENT

Per Pair of Students

- Incubated BHI agar plates from Period 2
- Incubated MSA plates from Period 2
- Incubated BHI tubes from Period 2
- Four tubes, each containing 1 ml reconstituted coagulase plasma EDTA medium (plasma tubes)
- Two toluidine blue DNA agar plates

Class Shared

- Incubator, set at 35°C
- Waterbath, set at 35°C
- Waterbath, set at 100°C

PROCEDURE

Inspecting BHI Agar and MSA Plates for Confirmation

1. Observe the *S. aureus* on BHI agar and MSA plates. Typical *S. aureus* colonies on BHI agar are round and yellow pigmented. *Staphylococcus aureus* colonies on MSA should be surrounded by yellow agar. This is indicative of mannitol fermentation.
2. Observe the *S. epidermidis* on BHI agar and MSA plates. Typical *S. epidermidis* colonies on BHI agar are round and off white in color. *Staphylococcus epidermidis* cannot ferment mannitol and *will not grow anaerobically on MSA*.
3. Observe the BHI agar and MSA plates streaked with typical *S. aureus* isolates from food. Determine whether or not the colonies match those of *S. aureus* (positive control). If matching, this confirms the *S. aureus* count which was reported earlier as typical (Table 7.2). If not, then that type of colony is not likely to be *S. aureus*.
4. Observe the BHI agar and MSA plates streaked with suspect isolates from food. Determine whether or not these colonies match the result of *S. aureus* (positive control). If matching, this confirms that the count which was reported earlier as suspect (Table 7.2) is indeed representing *S. aureus*. This count will

TABLE 7.3. *(Write a descriptive title, including data obtained and methods used)*

Isolate Number	Reaction on MSA	Appearance on BHI Agar	Confirmed *S. aureus* CFU/g
Typical			
1			
2			
3			
4			
Suspect			
1			
2			
3			
4			
Positive control			N/A
Negative control			

(Add appropriate footnotes clarifying descriptions and spelling out abbreviations).

then be included in the confirmed *S. aureus* count. If the suspect colonies did not produce the *S. aureus* typical reaction on the agar media, then the suspect count is not included in the total confirmed count.

5. Record these observations in Table 7.3.

Coagulase Test

1. Label two plasma tubes as "confirmed 1" and "confirmed 2."
2. Using an inoculation loop, transfer two to four colonies from BHI agar plates receiving confirmed *S. aureus* isolates into each of the plasma tubes. Use the loop to disperse the transferred colonies into the plasma medium.
3. Transfer two to four colonies of the positive control (*S. aureus*) from the BHI agar into a plasma tube, as indicated earlier. This tube is labeled "positive control."

TABLE 7.4. Coagulase and Thermonuclease Reaction Results of Two Confirmed *Staphylococcus aureus* Isolates and Positive and Negative Control

Isolate	Coagulase[a]	Thermonuclease
Confirmed 1		
Confirmed 2		
Positive control		
Negative control		

[a] Tubes were observed every (xx) min and final results were obtained after (x) hr of incubation.

4. Transfer two to four colonies of the negative control (*S. epidermidis*) from the BHI agar into a plasma tube, as indicated earlier. This tube is labeled "negative control."
5. Place the plasma tubes in the 35°C waterbath.
6. Periodically observe the plasma tubes.
   a. Gently tip the tube to a 45° angle.
   b. Coagulation (clotting or thickening) of the plasma indicates that the organism produces coagulase. It may take up to 6 hr for a positive reaction to develop.
7. Watch the positive control as an indicator of the progress of the reaction.
8. Record the result of coagulase testing in Table 7.4.

Thermonuclease Test

1. Place two incubated BHI tubes, receiving confirmed colonies, into the 100°C waterbath for 15 min, using extreme caution to prevent burns.
2. Draw a line that divides a toluidine blue DNA agar plate into two halves. Label these halves as "confirmed 1" and "confirmed 2."
3. Using an inoculation loop, streak the contents of the two heated sample tubes (step 1), each as a single straight line, onto the labeled toluidine blue DNA agar plate.
4. Draw a line that divides the second toluidine blue DNA agar plate into two halves. Label these halves as "positive control" and "negative control."
5. Using an inoculation loop, streak the contents of the heated positive control tube (step 1) onto the half of the toluidine blue DNA agar plate that is labeled "positive control." The streak should be a single straight line.
6. Streak the negative control onto the second half of the toluidine blue DNA agar plate, as indicated for the positive control.
7. Incubate the streaked toluidine blue DNA agar plates at 35°C for 48 hr.

Period 4 Data Collection

MATERIALS AND EQUIPMENT

Per Pair of Students

- Incubated toluidine blue DNA agar plate

Class Shared

- Colony counter

PROCEDURE

1. Examine the growth resulting from streaking the positive control onto toluidine blue DNA agar. Observe a pink halo around the streak (thermonuclease-positive reaction).
2. Examine the growth resulting from streaking the negative control onto toluidine blue DNA agar. Notice the lack of a pink halo and that the color around the streak remains blue (thermonuclease-negative).
3. Examine the growth resulting from streaking the confirmed colonies onto toluidine blue DNA agar. Determine the thermonuclease reaction of these two isolates.
4. Record the thermonuclease reaction of the four cultures in Table 7.4.

PROBLEMS

1. Why is there a need to enumerate, not just detect, *S. aureus* in food?

2. What type of food is considered favorable for growth and survival of *S. aureus* but considered unfavorable for other microorganisms? List at least two such foods.

3. Low count (e.g., 10^2–10^3 CFU/g) of *S. aureus* in a food indicates no potential *S. aureus*–associated health hazard in this food. Refute or substantiate this statement. Give evidence to support your stance.

4. For the food tested in this laboratory by you or your group, list the following information:
   (a) Method of packaging
   (b) Storage method
   (c) Sell-by date
   (d) Preservatives (if any)

5. Fill in Table 7.2 using your raw group data with proper footnotes.

6. Calculate and report the count of suspect *S. aureus* in food samples your group analyzed.

7. Fill in Table 7.3 using your group data with proper title and footnotes.

8. Report the confirmed count for *S. aureus*. Report the results of the coagulase and thermonuclease testing (Table 7.4).

9. The Baird–Parker medium used in this experiment is considered a selective–differential medium. Why?

10. How can processing or storage conditions affect the load of *S. aureus* in a food?

11. Compare the results of the Baird–Parker, MSA, and coagulase and thermonuclease tests. How accurate are the counts of suspect *S. aureus* in foods analyzed by your group, considering the results of the tests run to confirm the identity of the isolates?

12. Compare counts that the group obtained with the class average for the same food.

13. Compare and contrast counts between the foods that the class tested. Explain differences in the counts. Contamination is not a good answer unless you can explain why one food would be more contaminated than the other.

14. One can choose between testing for the bacterium (i.e., *S. aureus*) or the staphylococcal toxin it produces. Which test would you advise a company to do? Why?

15. Based on results obtained in this laboratory exercise, comment on the microbiological quality of the food analyzed. Are the data gathered sufficient to make a judgment about the suitability of this food for consumption?

CHAPTER 8

Listeria monocytogenes
USDA DETECTION PROTOCOL;
GENETIC IDENTIFICATION BY PCR TECHNIQUE

INTRODUCTION

Properties

The genus *Listeria* includes six species with only *Listeria monocytogenes* recognized as a human pathogen. These are non-spore-forming, gram-positive rods. All *Listeria* spp. produce catalase, lack oxidase, and hydrolyze esculin. *Listeria* spp. are differentiated based on carbohydrate fermentation and blood hemolysis (Table 8.1). Fermentation of xylose, rhamnose, and mannitol and results of hemolysis from blood agar stabs and a Christie–Atkins–Munch–Peterson (CAMP) test are useful in speciating *Listeria* isolates.

Listeria monocytogenes is a short rod and motile by a few flagella when cultured at 20–25°C. Motility is weak or lacking if the culture is incubated at 37°C. The characteristic tumbling motility can be observed by microscopic examination of a hanging-drop preparation. Overnight incubation of *L. monocytogenes* on nutritious agar medium produces colonies 0.2–0.8 mm in diameter. The colonies are smooth with entire margins and are translucent and bluish gray in color. When these colonies are examined using a suitable light source with an incidence light angle of 45° and a viewing angle of 135°, the typical blue-green iridescence can be seen.

Listeria monocytogenes grows well under aerobic and microaerophilic conditions. When grown on blood agar, *L. monocytogenes* shows β-hemolysis, although it may only be a narrow zone. The bacterium is homofermentative with glucose metabolized aerobically into lactate, acetate, and acetoin. Temperature for growth ranges from 1 to 45°C with the optimum at 35–37°C. The bacterium grows slowly at refrigeration temperature; generation time in skim milk at ~4°C is 30–40 hr. This ability of the microorganism to grow under refrigeration is used as a differential factor in

---

Food Microbiology   By Ahmed E. Yousef and Carolyn Carlstrom
ISBN 0-471-39105-0   Copyright © 2003 by John Wiley & Sons, Inc.

TABLE 8.1. Differentiation of *Listeria* spp. by Biochemical Testing and Blood Hemolysis

Species	Fermentation and Acid Production[a]			β-Hemolysis[b]	CAMP Test[c]
	Xylose	Rhamnose	Mannitol		
L. monocytogenes	−	+	−	+	+
L. innocua	−	v	−	−	−
L. seeligeri	+	−	−	+	+
L. welshimeri	+	v	−	−	−
L. ivanovii	+	−	−	+	−
L. grayi	−	−	+	−	−

[a] v = variable.
[b] When tested in agar media with sheep blood; see details about hemolysis later in this chapter.
[c] When tested against *Staphylococcus aureus*; see CAMP test details later in this chapter.

cold enrichment prior to isolation of the pathogen from food. The bacterium grows at relatively wide ranges of acidity (pH 4–9) and salt (up to 10%).

The Disease

Consumption of foods contaminated with *L. monocytogenes* may lead to an invasive infection and the resulting disease is known as listeriosis. It is generally believed that the infective dose is $>10^2$ listeriae; however, smaller doses are associated with the disease. The incubation period of listeriosis is one to several weeks. Children less than 4 years old, pregnant women, the elderly and immunocompromised individuals are the most susceptible to the disease. In pregnant women, the disease causes miscarriage, stillbirth, or premature birth. Listeriosis manifestation in susceptible nonpregnant adults includes bacteremia (symptoms resulting from the presence of the pathogen in the blood, e.g., fever, malaise, and fatigue), meningitis (inflammation of the membranes of the brain or the spinal cord), and meningoencephalitis (inflammation of the brain and its membranes). Symptoms associated with meningitis and meningoencephalitis include fever, malaise, seizures, and altered mental status. Mortality rate is 20–30%.

Recently, *L. monocytogenes* has been associated with a noninvasive disease called febrile gastroenteritis. The most common symptoms of this gastrointestinal listeriosis are fever and diarrhea, and the incubation period is only 20–27 hr. It appears that the infective dose of gastrointestinal listeriosis is much higher that it is in the invasive form of the disease.

Foods Implicated

Listeria monocytogenes is ubiquitous in the environment. Because of this, it has also been isolated from raw foods such as milk, salad vegetables, meat, poultry, and seafood. Processed ready-to-eat foods are occasionally contaminated with *L. monocytogenes*; these include frankfurters, sliced meats, and cheese. The pathogen also can be found in the food processing environment; hence, it is common that environmental samples are taken from processing facilities and tested for the presence of *Listeria* spp.

Detection

When *L. monocytogenes* is present in food, it is usually found at very low counts, and therefore it cannot be enumerated by direct plating. The infectious dose of *L. monocytogenes* is small; therefore, the mere presence of the pathogen in food constitutes a health hazard to consumers. Consequently, foods are commonly analyzed for the presence, rather than the number, of *L. monocytogenes*, and this analysis is described as "detection." Since a food sample weighing 25 g is analyzed, the theoretical minimum detection limit of the analysis is one *Listeria* per 25 g of food.

Conventional Detection Methods Typical methods for detection of *L. monocytogenes* include the following steps:

1. *Enrichment* Enrichment includes incubating a mixture of the sample in a rich microbiological medium that has limited or no selective properties. During incubation, injured *L. monocytogenes* recovers and the population of the pathogen increases. Cold enrichment was common decades ago, and this method relies on the ability of *L. monocytogenes* to grow at refrigeration temperature whereas the growth of other contaminants in the sample is arrested. Although cold enrichment gives good recovery of small *Listeria* populations, the method is time consuming as it takes weeks of refrigeration to produce satisfactory recovery of the pathogen. Alternatively, newer two-stage enrichment methods are currently used. This includes preliminary enrichment in a marginally selective broth followed by a second enrichment in a more selective medium.

2. *Isolation* Selective–differential agar media are traditionally used for isolation of *L. monocytogenes* from the enrichments. The most commonly used media for isolation of *L. monocytogenes* are Oxford; modified Oxford (MOX); and polymyxin, acriflavin, lithium chloride, ceftazidine, aesculin, and mannitol (PALCAM) agar media. These are highly selective media which contain salt (i.e., LiCl), antibiotics (e.g., polymyxin), and other antimicrobial agents that are tolerated by *Listeria* spp. but not by many other microorganisms. Isolation of *Listeria* from these two media is facilitated by addition of esculin and a ferric salt as differential agents. All *Listeria* spp. hydrolyze esculin into 6,7-dihydroxycoumarin, which reacts with ferric ions producing a black end product. McBride *Listeria* agar was commonly used for isolation of *Listeria* spp., but it has less selectivity than Oxford or PALCAM media.

3. *Identification* Suspect isolates from the selective–differential agar media may be identified as *L. monocytogenes* after biochemical characterization and confirmation. Acid production from rhamnose but not from xylose or mannitol is used in differentiating *L. monocytogenes* from other *Listeria* spp. Catalase reaction and blood hemolysis (e.g., CAMP test using *Staphylococcus aureus*) are also useful biochemical characterization tests. Subculturing these isolates on nonselective agar medium [e.g., tryptic soy agar–yeast extract (TSAYE) agar] should allow the analyst to detect the semitransparent blue-green or blue-gray appearance of the *L. monocytogenes* colonies. Tumbling motility can be examined microscopically in broth cultures which have been incubated at 20–25°C for 24 hr; higher incubation temperatures and longer incubation periods impair the motility of this pathogen. It is always a good

practice to Gram stain isolates and examine them microscopically. In a 24-hr culture, *L. monocytogenes* should produce gram-positive short rods. Older cultures may produce coccoidal and gram-variable cells.

Rapid Detection Methods Alternative rapid methods for detection of *L. monocytogenes* include an enrichment technique followed by an immunological or a genetic identification test. Therefore, the time-consuming isolation and biochemical characterization steps may not be needed when these alternative rapid methods are followed. Although negative results (i.e., absence of the pathogen) by rapid methods are considered conclusive, positive results by these methods need confirmation by the traditional cultural and biochemical techniques.

OBJECTIVES

1. Detect *L. monocytogenes* in food and the food processing environment.
2. Learn and practice a molecular identification technique in addition to cultural and biochemical methods.
3. Apply safe procedure for handling pathogens in the laboratory.

ORGANIZATION

Two laboratory exercises to detect *Listeria* in food are presented in this chapter. During the first exercise, students analyze food using a conventional method, whereas a rapid detection method is used in the second exercise (Table 8.2). The

TABLE 8.2. Detection of *Listeria* spp. in Foods and Environmental Samples Using Conventional and Rapid Methods

Period	Cultural/Biochemical Method		Genetic-Based Method	
	Step	Description	Step	Description
1	Primary enrichment	Culturing in UVM1	Primary enrichment	Culturing in UVM1
2	Selective enrichment	Subculturing in Fraser broth	Selective enrichment	Subculturing in Fraser broth
3	Isolation	Plating enrichments on • MOX agar • PALCAM agar	Identification (PCR based)	• DNA extraction • Amplification of DNA by PCR
4	Identification	Plating isolates on • TSAYE • TSA–blood	Identification (continued)	• Gel electrophoresis • Results interpretation
5	Identification (continued)	• Colony morphology • Cell morphology • Catalase reaction • Hemolysis		

conventional method is completed in five laboratory periods and the rapid method requires four periods. These two exercises may be run sequentially or simultaneously. Since the two methods share common enrichment steps, it saves time to run both exercises side by side. Procedural details of each exercise are presented later in the chapter.

PERSONAL SAFETY

Listeria monocytogenes is a highly pathogenic microorganism that should be handled with care. Follow the safety guidelines that were reviewed at the beginning of this manual. Use disposable gloves when handling isolates and the pathogenic cultures. Make sure the work area is sanitized after use. Pregnant women and immunocompromised individuals should not be allowed in the laboratory or areas where analysis for *L. monocytogenes* is in progress.

REFERENCES

Difco. 1998. *Difco Manual*, 11th ed. Difco Laboratories, Sparks, MD.

Food Safety Inspection Service. 1998. *Microbiology Laboratory Guidebook*, 3rd ed. U.S. Department of Agriculture, Washington, DC.

Jay, J. M. 2000. *Modern Food Microbiology*. Aspen Publishers, Gaithersburg, MD.

Lou, Y., and A. E. Yousef. 1999. Characteristics of *Listeria monocytogenes* Important to Food Processors. In E. T. Ryser and E. H. Marth (Eds.), *Listeria, Listeriosis and Food Safety* (pp. 131–224). Marcel Dekker, New York.

Ryser, E. T., and C. W. Donnelly. 2001. *Listeria*. In F. P. Downes and K. Ito (Eds.), *Compendium of Methods for the Microbiological Examination of Foods*, 4th ed. (pp. 343–356). American Public Health Association, Washington, DC.

EXERCISE 1: CULTURAL/BIOCHEMICAL METHOD (Modified USDA Protocol)

PROCEDURE OVERVIEW

Methods developed by researchers at the Food and Drug Administration (FDA) or the U.S. Department of Agriculture (USDA) are commonly used in the United States for the detection of *L. monocytogenes* in food. These methods rely mainly on the cultural and biochemical techniques. A modified and simplified version of the USDA protocol will be practiced in this laboratory exercise. The modified procedure allows students to detect and identify *Listeria* spp., but not *L. monocytogenes*, in the food and environmental samples. For identification of *L. monocytogenes*, additional biochemical testing is needed (Table 8.1). An overview of the method used in this laboratory exercise is shown in Table 8.2 and Fig. 8.1. The following are the main steps of this method.

Primary Enrichment

The initial enrichment step includes incubating a mixture of the food sample in a rich microbiological medium that has some selective properties. In this laboratory, the modified University of Vermont (UVM1) broth is used for this purpose. Samples of food and environmental swabs will be mixed with UVM1 and incubated. This step will be applied to the environmental sample (from a previous exercise; see Chapter 6) even though this sample has already been subjected to an enrichment during its cold storage.

Selective Enrichment

For selective enrichment, samples that have been preenriched for *L. monocytogenes* are mixed with a rich selective–differential broth (Fraser broth). The mixture is incubated to allow selective proliferation of *L. monocytogenes*. Darkening of the medium color may indicate the presence of *L. monocytogenes*.

Isolation

Samples enriched in Fraser broth are streaked on two highly selective–differential agar media, MOX agar and PALCAM agar media. The presence of characteristic *Listeria* colonies indicates successful isolation of the pathogen.

Identification

Four presumptive *Listeria* colonies from MOX and PALCAM are streaked onto TSAYE agar medium, plates are incubated, and isolates are examined further. Identification of these isolates includes morphological and biochemical testing. For morphological testing, isolates are examined on TSAYE for typical *Listeria* blue-green iridescence. Additionally, isolates are Gram stained and examined microscopically. Biochemical testing in this laboratory exercise will be limited to catalase reaction and β-hemolysis. Therefore, colonies from the TSAYE plates will be used for these tests.

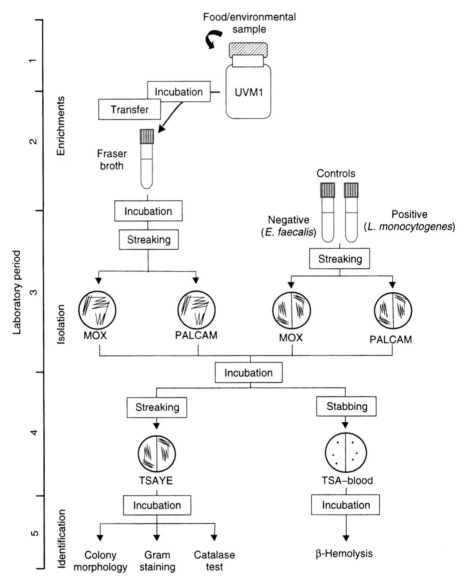

Fig. 8.1. Detection of *L. monocytogenes* in food and environmental samples using the cultural/biochemical method.

Data Interpretation

The cultural/biochemical method used in this laboratory exercise will be considered the reference protocol for detection of *Listeria* spp., and the results of the genetic-based method will be only confirmatory. When most analyses of the cultural/biochemical method produce results that are consistent with the typical characteristics of *L. monocytogenes*, a final conclusive result may be drawn: *Listeria* sp. is present in food or *Listeria* sp. is absent from food. Notice that there are cultural and biochemical variations among strains of *L. monocytogenes*, and thus certain "typical" reactions may not be seen with some isolates of the pathogen. Positive β-hemolysis, for example, is a typical reaction of *L. monocytogenes*, but some strains are weaker than others in lysing the red blood cells. In fact, a few strains of *L. monocytogenes* are nonhemolytic. Since a distinction between *L. monocytogenes* and other *Listeria* spp. require additional biochemical characterization, which is not done in this laboratory exercise, results of this analysis may be conclusive at the genus but not the species level.

MEDIA

Fraser Broth

In this exercise, Fraser broth is used in the selective enrichment step. Selectivity of this medium is due to the combined presence of acriflavin, nalidixic acid, and lithium chloride. Lithium chloride inhibits the growth of enterococci while permitting the growth of *Listeria* spp., which are salt tolerant. Esculin and ferric ammonium citrate are differential agents. *Listeria* spp. form a black precipitate in the presence of esculin and ferric ammonium citrate due to the reaction of an esculin hydrolysis product made by the bacteria with the iron in the ferric ammonium citrate. This reaction is responsible for the darkening of the broth. Thus, media with presumptive *Listeria* are dark, whereas negative tubes remain yellowish.

UVM1 Broth

This is a rich medium with some selectivity for *Listeria* spp. and is used in the preliminary enrichment step. The medium contains nalidixic acid, which inhibits most gram-negative bacteria. Acriflavin suppresses the growth of most other gram-positive bacteria. This medium is less selective than Fraser broth (which is also an enrichment broth), in order to allow the repair of injured cells. Although the medium contains esculin, it lacks soluble ferric salts and thus has no differential agents (compare with Fraser broth).

Modified Oxford (MOX) Agar

This is a selective–differential medium used for the isolation of *Listeria* spp. Selectivity of this medium is due to the presence of lithium chloride, colistin sulfate, and moxalactam. *Listeria* species are able to grow in the presence of these antimicrobiotics. Esculin and ferric ammonium citrate act together as differential agents. *Listeria* spp. produce an enzyme that hydrolyzes the esculin. The resulting hydrolysis product reacts with the iron (ferric) ions of the ferric ammonium citrate. This reaction produces a characteristic black precipitate. Lithium chloride inhibits the growth of gram-negative organisms while permitting the growth of *Listeria* spp., which are salt tolerant.

PALCAM Agar Medium

Similar to MOX, PALCAM is also a selective–differential medium used for the isolation of *Listeria* species. Selectivity of this medium is due to the presence of lithium chloride, polymyxin, acriflavine, and ceftazidime. Differential agents in this medium are esculin–ferric ammonium citrate and mannitol–phenol red combinations. Hydrolysis of esculin by *Listeria* leads to medium blackening (see previous explanation), and growth of mannitol-fermenting bacteria (e.g., some staphylococci and enterococci) will acidify the medium and changes its color from red or gray to yellow.

Tryptic Soy Agar with Blood (TSA–Blood)

This medium is used for cultivating fastidious microorganisms and for differentiating those causing blood hemolysis. The medium is made of TSA with addition of ~5% sheep blood. To prepare TSA–blood, melt TSA agar and cool to 50°C. To 1 liter of molten TSA, add 50 ml defibrinated sheep blood and mix well without incorporating air bubbles in the mixture. Poor the mixture thickly in Petri plates (~20 ml/plate); the agar should appear bright red, opaque, and firm. Leave the poured plates at room temperature (~22°C) for 48 hr before use; this allows proper drying of agar surface. To produce optimum results, *L. monocytogenes* is stabbed into TSA–blood and plates are incubated at 35°C for 48 hr, preferably under increased CO_2 atmosphere. Under these conditions, *L. monocytogenes* produced distinct but narrow zones of β-hemolysis around the stabs. Possible types of hemolysis on blood agar media, in general, are as follows:

- α-*Hemolysis*: Reduction of the red blood color to a greenish discoloration around the colony. This results from conversion of blood hemoglobin to methemoglobin by the hemolytic microorganism.
- β-*Hemolysis*: Production of a clear zone around the colonies due to the lysis of red blood cells by the hemolytic microorganisms.
- γ-*Hemolysis*: This indicates no hemolysis and no change in the color of blood around the colonies.

Tryptic Soy Agar with Yeast Extract (TSAYE)

This medium is made of TSA with addition of 2% yeast extract. The medium is a relatively rich, nonselective medium. The medium is used in this experiment to provide a clear background that allows examination of colony morphology since the medium does not interfere with the appearance of the characteristic blue-gray tint of *Listeria* colonies. Colonies growing on this medium also are suitable for testing the catalase reaction.

ORGANIZATION

These exercises require five laboratory periods to complete (Table 8.2). Students will work in pairs with one student analyzing the food sample and the other analyzing the environmental sample. Possible foods include meat, seafood, and raw vegetables. The group members should choose one of the environmental samples that were collected and cold enriched in a previous laboratory exercise (Chapter 6). Group members should exchange experience and share results. When running the polymerase chain reaction (PCR) technique, students will work in groups of four.

Period 1 Enrichment

MATERIALS AND EQUIPMENT

Per Pair of Students

- Food sample
- Environmental sample from a previous laboratory exercise (Chapter 6)
- UVM1 broth:
  (a) For food sample: 225 ml
  (b) For environmental sample: 90 ml
- Stomacher bags (alternatively, use sterile jars with lids)
- Large beakers or tubs for holding stomacher bags during incubation

Class Shared

- Scale for weighing food samples (e.g., a top-loading balance with 500 g capacity)
- Stomacher
- Incubator, set at 30°C

PROCEDURE

Food Sample

1. Prepare the food for sampling to ensure that a representative sample is analyzed. This varies with the food and may involve cutting, partitioning, grinding, or mixing.
2. Weigh 25 g of the food sample into a stomacher bag. Add the 225 ml of UVM1 broth to the bag contents. Stomach for 2 min.
3. Place the bag in the beaker designated by the laboratory instructor for holding the enrichment bags. This container should support the bag and keep it upright during incubation and handling. Alternatively, transfer the bag contents into a sterile jar and close with the lid. (*Note: The jar should be autoclaveable so that its contents can be heat sterilized after use.*)
4. Incubate the enrichment bags or jars at 30°C for 24 hr.

Environmental Sample

1. Transfer the entire contents of the environmental sample tube (including the swab) into the 90-ml flask of UVM1. The storage of this sample at 4°C has served as a cold enrichment.
2. Incubate the flasks at 30°C for 24 hr.

Period 2 Selective Enrichment

MATERIALS AND EQUIPMENT

Per Pair of Students

- Enriched environmental sample
- Enriched food sample
- Two 9-ml tubes of Fraser broth
- Latex gloves

Class Shared

- Incubator, set at 35°C

PROCEDURE

1. Mix the culture by hand agitating the stomacher bag or swirling the flask.
2. Transfer 1 ml of the enriched food sample into a labeled tube of Fraser broth. Repeat for the environmental sample.
3. Incubate the inoculated Fraser broth tubes at 35°C for 24 hr. Alternatively, incubate at room temperature (~22°C) for 48 hr.

Period 3 Isolation

MATERIALS AND EQUIPMENT

Per Pair of Students

- Latex gloves
- Environmental sample enriched in Fraser broth (environmental sample enrichment)
- Food sample enriched in Fraser broth (food sample enrichment)
- Seven plates of MOX agar medium (three for food sample, three for environmental sample, one for positive and negative controls)
- Seven plates of PALCAM agar medium (three for food sample, three for environmental sample, one for positive and negative controls)
- One tube of *L. monocytogenes* overnight culture (positive control)
- One tube of *Enterococcus faecalis* overnight culture (negative control)

Class Shared

- Incubator, set at 35°C

PROCEDURE

1. Label plates: Each set of seven includes three plates for the food sample, three for the environmental sample, and one for the positive and negative controls. Draw a line that divides each of the control plates into halves; one for the positive and the other for the negative control.
2. Three-phase streak the *food sample* enrichment onto three plates of each of the MOX and the PALCAM agar media.
3. Three-phase streak the *environmental sample* enrichment onto three plates of each of the MOX and the PALCAM agar media.
4. Two-phase streak the *positive-control* culture (*L. monocytogenes*) and the *negative-control* culture (*E. faecalis*) onto the MOX and the PALCAM agar plates.
5. Incubate the plates at 35°C for 24 hr. Alternatively, incubate at room temperature (~22°C) for 48 hr.

Period 4 Identification

MATERIALS AND EQUIPMENT

Per Pair of Students

- Incubated MOX and PALCAM agar plates (prepared during the previous laboratory period)
- Three TSAYE agar plates (for food sample, environmental sample, and controls; one plate each)
- Three TSA–blood agar plates, thickly poured (one each for food and environmental samples, the other for controls)
- One tube of *E. faecalis* overnight culture (negative control)
- Latex gloves

Class Shared

- Colony counter

PROCEDURE

Isolation (Continued)—Examination of Agar Plates

Examine the colonies on the selective–differential agar plates prepared in the previous laboratory period. Colonies with typical reactions on these media are considered presumptive *Listeria* isolates.

MOX Agar Medium

1. Examine the appearance of the colonies on the control plate. Colonies of *Listeria* are differentiated by their dark color surrounded by a black zone due to the hydrolysis of esculin. Examine the negative control for any detectable growth.
2. Examine the food and environmental sample plates for any presumptive *Listeria* colonies. If present, mark two isolated colonies that have the characteristic *Listeria* appearance; one from the food source and the other from the environmental sample.
3. Record the observations in Table 8.3.

PALCAM Agar Medium

1. Examine the appearance of the colonies on the control plate. Colonies of *Listeria* appear dark and are surrounded by a blackened agar. Additionally, *L. monocytogenes* cannot ferment mannitol. Mannitol-fermenting bacteria may grow on this agar, and their colonies are surrounded by yellow-colored agar due to acid production. Examine the negative control for any detectable growth.

TABLE 8.3. Isolation of Presumptive *Listeria monocytogenes* Colonies on Selective Agar Media and Biochemical and Morphological Characterization of These Isolates

Source	Presence or Absence of Presumptive Colonies and Colony Descriptions			Gram Stain	Catalase Test	Hemolysis
	MOX Agar	PALCAM Agar	TSAYE			
Food sample						
Environmental sample						
L. monocytogenes				$+^a$ (short rod)	$+^b$	$+^c$
E. faecalis				+ (coccus)	−	−

[a] Gram-positive reaction.
[b] Positive: bubbles formed.
[c] β-Hemolysis.

2. Examine the food and environmental sample plates for any presumptive *Listeria* colonies. If present, mark two isolated colonies that have the characteristic *Listeria* appearance: one from the food source and the other from the environmental sample.
3. Record the observations in Table 8.3.

Subculturing for Identification

TSAYE Agar Medium
Controls

1. Label a TSAYE agar plate and draw a line that divides the plate into two equal divisions.
2. Using inoculation loop, pick a colony from the incubated positive controls (MOX or PALCAM agar plates) and two-phase streak onto the positive-control half of the TSAYE agar plate.
3. Repeat the previous step using the negative control. If no negative-control cultures grew on MOX or PALCAM agar plates, use an overnight-incubated negative-control culture broth.

Isolates from Food Sample

1. Label a TSAYE agar plate and draw a line that divides the plate into two equal divisions.
2. Using an inoculation loop, pick a portion of the marked presumptive *Listeria* colony from the incubated food sample–MOX agar plate. Streak this isolate (two-phase streaking) onto one-half of the TSAYE agar plate.

3. Repeat the previous step with the marked colony on food sample–PALCAM agar plates and streak on the other half of the TSAYE agar plate.

Isolates from Environmental Sample

1. Label a TSAYE agar plate and draw a line that divides the plate into two equal divisions.
2. Using an inoculation loop, pick a portion of the marked presumptive *Listeria* colony from the incubated environmental sample–MOX agar plate. Streak this isolate (two-phase streaking) onto one-half of the TSAYE agar plate.
3. Repeat the previous step with the marked colony on environmental sample–PALCAM agar plates and streak on the other half of the TSAYE agar plate.

All TSAYE Plates

1. Incubate the TSAYE agar plates at 35°C for 24 hr. Alternatively, incubate at room temperature (~22°C) for 48 hr.
2. If no presumptive *Listeria* colonies were detected on MOX or PALCAM agar plates, then testing is complete and a negative result should be recorded.

TSA–Blood Medium
Controls

1. Label a TSA–blood plate and draw a line that divides the plate into two equal divisions.
2. Using an inoculation needle, pick a colony from the incubated positive controls (MOX or PALCAM agar plates) and stab the agar several times in the positive-control half of the TSA–blood plate.
3. Repeat the previous step using the negative control. If no negative-control cultures grew on MOX or PALCAM agar plates, use an overnight negative-control culture.

Isolates from Food Sample

1. Label a TSA–blood plate and draw a line that divides the plate into two equal divisions.
2. Using an inoculation needle, pick a portion of the marked presumptive *Listeria* colony from the incubated food sample–MOX agar plate. Stab the agar several times with the inoculum using one-half of the TSA–blood plate.
3. Repeat the previous step with the marked colony from food sample–PALCAM agar plates and stab the other half of the TSA–blood agar plate.

Isolates from Environmental Sample

1. Label a TSA–blood plate and draw a line that divides the plate into two equal divisions.
2. Using an inoculation needle, pick a portion of the marked presumptive *Listeria* colony from the incubated environmental sample–MOX agar plate. Stab the agar several times with the inoculum using one-half of the TSA–blood plate.
3. Repeat the previous step with the marked colony from environmental sample–PALCAM agar plates and stab the other half of the TSA–blood agar plate.

All TSA–Blood Plates

1. Incubates the TSA–blood agar plates at 35°C for 48 hr.
2. If no presumptive *Listeria* colonies were detected on MOX or PALCAM agar plates, then testing is complete and a negative result should be recorded.

Period 5 Identification (Morphological and Biochemical Characterization)

MATERIALS AND EQUIPMENT

Per Pair of Students
- Gram-staining reagents
- Hydrogen peroxide, 6% solution
- Incubated TSAYE agar plates
- Incubated TSA–blood agar plates
- Microscope
- Light source for colony morphology examination

PROCEDURE

TSAYE Agar Plates

Colony Morphology
1. Examine the appearance of the colonies on the control TSAYE plate using an obliquely positioned light source. Typical *Listeria* colonies are 0.2–1.5 mm in diameter and appear blue-gray when seen using a 45° (oblique) light setup. To see the coloration, tilt the plate while holding it over the light source.
2. Examine the food and environmental sample plates. Mark isolated colonies that have the characteristic *Listeria* appearance.
3. Record the observations in Table 8.3.

Cell Morphology
1. Pick a colony of *L. monocytogenes* from the TSAYE control plate and Gram stain. Follow the Gram-staining technique as indicated in Chapter 1.
2. If presumptive *Listeria* colonies are observed on the TSAYE agar plate receiving food samples, pick a colony using the inoculation loop and perform a Gram staining.
3. Repeat the former step using TSAYE plates receiving presumptive isolates from the environmental sample.
4. Examine stained smears under the microscope (see Chapter 1) and determine isolate Gram reaction and cell morphology.
5. If no presumptive *Listeria* were cultured on TSAYE, then Gram stain the positive and negative controls.
6. Record the observations in Table 8.3.

Catalase Test
1. Perform a catalase test on a colony from the positive control, the negative control, the food sample, and the environmental sample.

TABLE 8.4. Results of Detection of *Listeria* Species in Food and Environmental Samples

Source	Modified USDA Method	Genetic Rapid Method	Remarks
Food sample (_____)			
Environment sample (_____)			

2. Transfer a portion of the colony to a microscope slide and add a single drop of 6% hydrogen peroxide solution. Instant effervescence due to the production of oxygen bubbles indicates the presence of the catalase enzyme. No bubbles is a negative reaction.
3. Record the observations in Table 8.3.

TSA–Blood Plate (Blood Hemolysis)

1. Examine the TSA–blood plates receiving stabs from the control, the food sample, and the environmental sample.
2. Observe any blood clearing around the stabs. Use the light source to observe the narrow zone of hemolysis, if present.
3. Record the hemolysis results in Table 8.3.

Evaluation of Results

1. Evaluate the results of the food and environmental samples as recorded in Table 8.3. Determine if the sample is positive or negative for the presence of *Listeria* spp. or whether the result is inconclusive (see the Procedure Overview section for data interpretation). Record the final results of this analysis in Table 8.4.
2. Keep in mind that positive samples, as determined in this exercise, would be subjected to additional tests (e.g., motility, biochemical and serological tests) to confirm if the isolates are *L. monocytogenes*.
3. The CAMP test is a possible confirmatory test, particularly when hemolysis around the stabs in TSA–blood medium is questionable. To run this test, *L. monocytogenes* is streaked on blood agar in proximity (and perpendicular) to a streak of a β-hemolytic *S. aureus*, and the plate is incubated 24–48 hr at 35°C. By carefully examining the plate using a colony counter, an enhanced zone of hemolysis is observed where these streaks are closest to each other. Observe the CAMP test demonstration plate prepared by the laboratory instructor. Compare the results of the positive and negative controls.

EXERCISE 2: GENETIC-BASED RAPID METHOD

A rapid alternative, genetic-based detection method may be run simultaneously with the cultural/biochemical protocol. The genetic method chosen for this exercise includes a PCR-based identification technique substituting the conventional isolation and identification steps in the cultural/biochemical method. An overview of the genetic-based method is shown in Fig. 8.2 and Table 8.2. Although this method is less time consuming (than the conventional methods) in laboratories where these analyses are run routinely, this time saving is less apparent in most teaching laboratories.

PROCEDURE OVERVIEW

A commercial test kit (BAX System, Qualicon, Wilmington, DE) that is specific for the genus *Listeria* (i.e., *Listeria* spp.) will be used. The kit includes a lysis reagent for extraction of *Listeria* genomic DNA from the enriched sample, a *Listeria*-specific DNA probe, a nonspecific probe that functions as a control, the polymerase enzyme, and the needed nucleotides. Since there have been frequent procedural modifications by this commercial kit producer, it is recommended that method updates are sought directly from the kit's manufacturer.

Enrichment

The genetic-based method consists of primary and selective enrichment steps similar to those used in the cultural/biochemical method. If the two methods are run simultaneously, the sample enrichments can be shared.

Cell Lysis

A portion of incubated Fraser enrichment broth is mixed with a lysis reagent (included in the BAX kit) and the mixture is heated at 55 and 95°C to cause cell lysis and release of genomic DNA.

DNA Amplification (PCR Technique)

The lysed cells are transferred to a PCR tube which contains a reagent pellet (included in the BAX kit). The pellet includes a *Listeria*-specific DNA probe, a nonspecific probe (control), Taq polymerase, and nucleotides. The tubes are transferred to the thermocycler, which has been preprogrammed to heat and cool in cycles. After 38 cycles of heating (94°C) and cooling (70°C), the reaction mixture is cooled to 25°C. The mixture-containing tubes may be kept overnight at 25°C in the PCR wells or transferred to a refrigerator at 4–6°C for a longer storage period.

DNA Detection (Gel Electrophoresis)

Each agarose gel is cast to contain 12 wells (enough for a group of four students running their samples in duplicates) and placed in the electrophoresis unit, with the

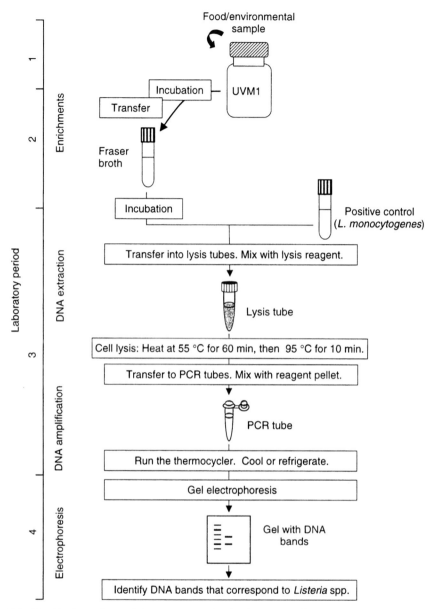

Fig. 8.2. Rapid detection of *Listeria* spp. in food and environmental samples using a genetic-based method.

wells toward the negative electrode. The agarose gel should be completely immersed in the running buffer before sample loading. The lysed samples are mixed with a loading (tracking) dye and the mixtures are dispensed in the agarose wells. The mass ladder is also loaded in one of the wells. Electrophoresis is performed by connecting the electrophoresis unit to electricity and adjusting and maintaining the voltage at the desired setting. Keeping track of samples during preparation, PCR, and gel electrophoresis is crucial for eliminating sample mixup.

Gel Photographing

The gel is stained with ethidium bromide unless this reagent was already included in the cast gel; the stain will help visualize the DNA bands under UV illumination. When the gel is exposed to UV light, the DNA bands will be visible and a picture is taken using a light or a digital camera. The analyst's body should not be exposed to the ethidium bromide or to the UV light and therefore protective glasses and gloves should be worn during this exercise.

Results Explanation

The sizes of the bands resulting from the sample will be determined by comparing their position relative to the bands of the mass ladder. According to the kit's manufacturer, a band of 500 base pairs (bp) is indicative of the presence of *Listeria* species. Presence of a 200-bp band (control) on the gel indicates proper functioning of the polymerase reaction. Under ideal conditions, both the 200- and 500-bp bands will be present when the sample contains *Listeria* spp. Presence of the 500-bp band but not the 200-bp band is also acceptable as a positive *Listeria* spp. result, since a large amount of target DNA may have competed with the control DNA for reagents during the PCR cycles. Absence of the 500-bp band when the 200-bp band is present indicates the sample is negative for *Listeria* spp. Lack of both 200- and 500-bp bands indicates test failure. Band sizes should be checked in the kit's instruction pamphlet to ensure proper interpretation.

Data Interpretation

If no *Listeria* spp. are detected in food by the genetic-based procedure, the result may be considered conclusive, particularly if the streaking on MOX and PALCAM agar media (during the cultural/biochemical procedure) also produced negative results. If DNA bands typical to *Listeria* spp. are found on the agarose gel, the result may be used to confirm similar positive results from the cultural/biochemical method.

Period 1 Enrichment

Use the same steps as in the cultural/biochemical method.

Period 2 Selective Enrichment

Use the same steps as in the cultural/biochemical method.

Period 3 Genetic Identification: Cell Lysis and PCR Reaction

MATERIALS AND EQUIPMENT

Per Group of Four Students

- Latex gloves
- Environmental sample enriched in Fraser broth (environmental sample enrichment)
- Food sample enriched in Fraser broth (food sample enrichment)
- Culture of *L. monocytogenes* (positive control)
- Five sterile 1-ml microcentrifuge tubes (sample tubes)
- Five sterile 1-ml screw-cap microcentrifuge tubes (lysis tubes). Screw-cap is essential to prevent the loss of reaction mixture when pressure builds up during heating.
- Floating microcentrifuge tube rack (fits lysis tubes)
- Five PCR tubes; these are ~50-µl conical tubes with snap caps
- Pipettors and tips (20, 200-µl capacities)
- Lysis reagent (prepared from ingredients that come with the BAX kit; made of lysis buffer and a protease)
- PCR tube labels
- Cell lysis tube rack
- Bucket of ice

Class Shared

- Thermocycler for PCR
- PCR tube holder
- Water bath, set at 55°C; alternatively, a heating block set at 55°C
- Water bath, set at 95°C; alternatively, a heating block set at 95°C

PROCEDURE

Sample Preparation for DNA Extraction

Students will be working in groups of four for this experiment, each pair with a food and an environmental sample and using a common positive control.

1. Label the sample tubes: two for the food sample enrichments, two for the environmental sample enrichments, and one for the positive control.
2. Mix the contents of the food sample enrichment tube and transfer 1 ml into appropriately labeled sample tube. [*Note:* It is easier and safer to pipette 5 µl enrichment (at a subsequent step) from this 1-ml tube than from the larger Fraser broth enrichment tube.]
3. Repeat the previous step using the second food sample enrichment, the two environmental sample enrichments, and the positive control.
4. Label the lysis tubes: two for the food sample enrichments, two for the environmental sample enrichments, and one for the positive control.
5. Dispense 200 µl lysis reagent into each of the lysis tubes.
6. Transfer 5 µl enrichment from a sample tube (one of the tubes containing the food sample enrichments) into the appropriately labeled lysis tube. Mix the tubes gently by tapping with a finger.
7. Repeat the previous step using the enrichments of the second food sample, the two environmental samples, and the positive control.

Cell Lysis

1. Place the lysis tubes in the floating rack, then place the rack in the 55°C waterbath and incubate for 60 min.
2. Transfer the rack with the lysis tubes to the 95°C waterbath for 10 min. Use heavy gloves (not latex) when opening the waterbath; the water is extremely hot.
3. Cool the lysis tubes on ice for at least 5 min before use in subsequent steps. Keep tubes on ice at all times unless they are being processed at subsequent steps.

Preparation for PCR Amplification

1. Using a small tag, label one PCR tube for each sample and for the control (a total of five tubes). Include group identification on the label. Be careful not to disturb or lose the PCR reaction pellet in the tubes during labeling.
2. Transfer 50 µl of the food sample lysis mixture (from the lysis tube) into the appropriately labeled PCR tube. Place the tube on ice.
3. Repeat the previous step for the second food sample, both environmental samples, and the positive control.

Amplification of Target DNA by PCR

1. Since writing comes off easily from the PCR tubes, check the labeling on the tubes; refresh the writing as required. Additionally, marking the location of the PCR tubes during the rest of the procedure should alleviate labeling confusion.
2. Place the PCR tubes in a suitable rack and record their location on a "diagram of the racks," prepared by the laboratory instructor.
3. Turn on the thermocycler and program it as follows:
   a. Initial holding: 94°C for 2 min
   b. Heating cycle: DNA denaturation at 94°C for 15 sec followed by primer binding and polymerase reaction at 70°C for 3 min
   c. Heating cycle repeats: 38 times
   d. Cooling: at 25°C
4. Place the PCR tubes in the thermocycler receptacle and label their location on the "diagram of the thermocycler wells," provided by the laboratory instructor.
5. Start the thermocycler.
6. When the heating cycles are completed and samples are cooled, return the samples to their original locations in the PCR tube racks and store the samples in the refrigerator for analysis during the subsequent period.

[*Note:* It is recommended that laboratory instructor prepare a blank (negative control) using peptone water instead of Fraser broth enrichment and run it into the thermocycler along with the class samples.]

Period 4 Genetic Identification: Gel Electrophoresis

MATERIALS AND EQUIPMENT

Per Pair of Students

- PCR tubes from previous period
- Precast gel containing ethidium bromide (nucleic acid grade agarose, 2%; 0.5X tris borate EDTA buffer; ethidium bromide, 0.22 µg/ml gel)
- Loading dye
- Ice buckets
- UV protective glasses
- Mass markers (DNA ladder)
- Blank solution (buffer solution or 0.1% peptone water)
- Latex gloves

Class Shared

- Gel electrophoresis equipment
- UV transilluminator
- Instant camera setup (alternatively, a digital camera mounted on the transilluminator is ideal for taking gel pictures and later distributing these pictures)

PROCEDURE

Sample Preparation

1. Retrieve PCR tubes and place them on ice.
2. Add 15 µl of loading dye to each PCR tube.

Loading the Gels

1. Determine in advance the lane location of each sample to be loaded; record these locations on a chart next to the gel box. Maintaining lane assignments will help each student pair keep track of their samples and aid the laboratory instructor in keeping track of gel box assignments.
2. Carefully load 15 µl from each sample and control PCR tube into the assigned individual wells in the agarose gel.
3. Load 6 µl of mass markers (DNA ladder).
4. Load 15 µl of the blank solution. (*Note:* It may be preferable if the last two steps are carried out by the laboratory instructor.)

Gel Processing

1. Run the gel at 130–150 volts for approximately 60 min.
2. When the loading dye moves two-thirds of the gel length, turn off the electricity.

3. Remove the gel from the box.
4. Place the gel on the UV transilluminator (light box) and observe the fluorescent bands. (*Note:* Protective glasses should be worn to protect the eye from UV light. Exposure of skin to UV should also be avoided.)
   a. Each group should observe their gel during this laboratory period as it is illuminated on the light box.
   b. Take a picture of the gel using an instant or a digital camera.

Period 4 (*Continued*) Electrophoresis Results Explanation

MATERIALS

- Photograph of gel

PROCEDURE

DNA Ladder

The DNA ladder preferred in this experiment has bands with lengths of 100–1200 bp at 100-bp intervals. The DNA ladder that comes with the BAX kit has a lesser number of bands and thus is not ideal for teaching purposes. Notice that the distances between the large-sized bands in the ladder are smaller than those between the low-molecular-weight bands. Linear DNA fragments migrate through the gel at a rate inversely proportional to the log of their molecular weight. By comparing the position of bands of known molecular weight (DNA ladder lane) to the position of bands in other lanes, it is possible to estimate weights (in base pairs) of the bands in the other lanes.

Blank

No bands should be visible in this lane. This is the negative control.

Bands Representing Two Primer Sets

Two primer sets were used in the PCR, each specific to a different DNA sequence. The band at 500 bp is a result of a *Listeria* spp. specific primer sequence. The band at 200 bp is a control sequence used in verifying proper functioning of the PCR process.

Observations and Results Recording

Observe the following and record the results of these observations:

TABLE 8.5. Interpretation of Band Pattern to Determine Whether Culture Is *Listeria* Positive

Band at 500 bp?	Band at 200 bp?	Interpretation
Yes	Yes	Positive
Yes	No	Positive
No	Yes	Negative
No	No	Test failed

1. Do the bands for the DNA marker (ladder) appear normal?
2. Is the band representing the kit control visible at the appropriate location on the gel?
3. Food sample: At what positions are the bands?
4. Environmental sample: At what positions are the bands?
5. *Listeria* sample: At what positions are the bands?

Results Explanation

Use Table 8.5 to interpret the gel electrophoresis results. See also the section on Procedure Overview for electrophoresis results explanation.

Period 4 (*Continued*) Data Interpretation

Record the final results of the analysis using both the cultural/biochemical and the genetic-based methods in Table 8.4. For data interpretation, see the section on Procedure Overview.

PROBLEMS

1. Why is it important to adhere to all safety precautions during food analyses for *L. monocytogenes*?

2. Why is the food industry more concerned about the presence or absence of *L. monocytogenes* than the actual count of the pathogen in a given food? Compare this attitude with that toward *S. aureus*.

3. The PCR technique is used as an alternative to the biochemical methods that are commonly used to identify *Listeria* isolates from food. Give two advantages and two disadvantages of using the PCR technique for this purpose.

4. For the food tested in this laboratory, list very briefly the following information:
   (a) Method of packaging
   (b) Storage method
   (c) Sell-by date

5. Answer the following in regard to the food analyzed in this laboratory exercise:
   (a) Does this food have a history of contamination with *L. monocytogenes*, as evidenced by frequent product recalls?
   (b) Has this type of food been implicated in disease outbreaks attributed to the presence of *L. monocytogenes*?

6. What location in processing facility was used to take the environmental sample that was analyzed in this laboratory exercise? Did this area look clean? What did the analysis of this environmental sample prove, in terms of the cleanliness of the area sampled? (Check the results of analysis done on this sample in Chapter 6.)

7. Fill in Table 8.3 with group data.

8. Attach the picture of the gel which was taken during this exercise (label it Fig. 8.3). Add a descriptive title to the figure and add any needed legend.

Fig. 8.3. _____.

9. Interpret the results obtained from Fig. 8.3.

10. Would you recommend the genetic-based method to replace the cultural/biochemical method when analyzing food for *Listeria* spp. or *L. monocytogenes*? Explain your answer.

11. Considering the class data (from this exercise and the environmental microbiota exercise), is there a correlation between the microbial load and presence of *Listeria* spp. in the environmental sample? Has this correlation been cited in published literature? If so, list one reference, at least, supporting or refuting this assumption.

12. Why is it more common in the food industry to analyze food for *Listeria* spp. than *L. monocytogenes*?

13. Complete the missing information in the accompanying figure (Fig. 8.4).

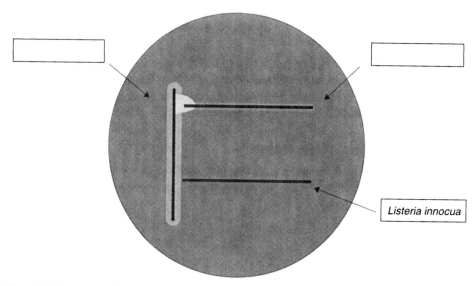

Fig. 8.4. Detection of *L. monocytogenes* (_____) activity and comparison to another microorganism on a CAMP test.

CHAPTER 9

Salmonella
FDA DETECTION PROTOCOL; ELISA TECHNIQUE;
DNA PROBE TECHNIQUE

INTRODUCTION

Properties

The genus *Salmonella* belongs to the family *Enterobacteriaceae*. As the name implies, *Enterobacteriaceae* is comprised of bacteria that proliferate in the intestine. Several genera of this family include pathogenic species. In addition to *Salmonella*, *Enterobacteriaceae* includes food-transmitted pathogenic species of these genera: *Escherichia*, *Shigella*, and *Yersinia*.

Like other *Enterobacteriaceae* genera, *Salmonella* is a gram-negative flagellated rod-shaped bacterium. Salmonellae are facultative anaerobes with both respiratory and fermentative metabolic pathways. They are oxidase negative, ferment glucose and produce acid and gas, grow on citrates as a sole source of energy, decarboxylate lysine and ornithine, generally produce hydrogen sulfide, and do not hydrolyze urea. One of the characteristics of this genus is that most members do not ferment lactose or sucrose.

Classification and Nomenclature

Based on their reaction with specific antibodies, *Salmonella* isolates can be classified into ~2400 *Salmonella* serovars. This classification is based on the type of antigens produced by isolates.

> *Somatic (O) Antigens* The O antigens are associated with the lipopolysaccharides (LPS) on the external surface of the bacterial outer membrane. Somatic antigens are heat stable and resistant to alcohol and dilute acids.

Food Microbiology By Ahmed E. Yousef and Carolyn Carlstrom
ISBN 0-471-39105-0 Copyright © 2003 by John Wiley & Sons, Inc.

Flagellar (H) Antigens These are the antigens associated with the peritrichous flagella. Flagellar antigens are heat-labile proteins.

Capsular (K) Antigens These antigens are produced by *Salmonella* serovars that produce capsular material. Capsular antigens are heat-sensitive carbohydrates.

Salmonella serovars cannot be differentiated biochemically. The Kauffmann–White scheme for classifying salmonellae assigns a species status to each serovar. Therefore, differentiation between *Salmonella typhimurium* and *S. enteritidis*, for example, is based on serotyping. According to a more recent classification scheme, serovars are not different species because they have ≥70% DNA homology, and numerical taxonomy supports their similarity. Therefore, two *Salmonella* species only are now recognized by the CDC: *Salmonella enterica* and *Salmonella bongori*. According to this scheme, the vast majority of the serovars are under *S. enterica*, with only 20 serovars belonging to *S. bongori*. Under this scheme, what used to be *S. typhimurium* is now *S. enterica* serovar Typhimurium or simply *Salmonella* Typhimurium. This latter scheme will be followed in this chapter.

The Disease

Salmonella is an invasive bacterium that causes human infection, known as salmonellosis. This disease is widely spread in many countries and affects the young and the elderly the most. Although it is assumed that *Salmonella* infectious dose is high, infection may result from ingesting as little as 1–10 cells.

There are different syndromes of human salmonellosis. Typhoid (enteric) fever is caused by *Salmonella* Typhi and *Salmonella* Paratyphi. It accounts for <5% of cases of salmonellosis. Symptoms include diarrhea, abdominal pain, headache, and prolonged high fever. The incubation period varies from 1 to 7 weeks, and disease may last 1–8 weeks. The most common form of human salmonellosis is the *Salmonella* gastroenteritis (or enterocolitis). This is caused by at least 150 serovars of *Salmonella*. Symptoms of *Salmonella* gastroenteritis include diarrhea, abdominal pain, chills, moderate fever, vomiting, dehydration, and headache. Incubation period is 12–36 hr, and duration is 1–4 days.

Foods Implicated

Since the intestine is the natural habitat of *Salmonella* serovars, raw foods of an animal source occasionally harbor the pathogen. *Salmonella*, therefore, is found in poultry products, including chicken, eggs, and turkey. Shellfish, milk, salads, and cantaloupes also have been implicated in salmonellosis outbreaks. Water has been a vehicle for transmission of salmonellosis. Workers who do not observe proper personal hygiene, especially those working in food harvesting, processing, and service, are potential sources of food contamination with *Salmonella*. The most common serovar-associated food-transmitted salmonellosis are caused by *Salmonella* Typhimurium and *Salmonella* Enteritidis.

Most *Salmonella* serovars grow only in the mesophilic temperature range; however, a few serovars can grow at temperatures as low as 2–4°C or as high as 54°C. Foods with pH < 4.5 do not normally support the growth of *Salmonella*, but some serovars grow at pH 4.0 (e.g., *Salmonella* Infantis). Pasteurization temperatures readily inactivates *Salmonella* serovars. One serovar (i.e., *Salmonella* Senftenberg 775W), at least, is known to have exceptional resistance to heat. Recent appearance of isolates with multidrug resistance (e.g., *Salmonella* Typhimurium DT 104) is a potential threat to the safety of consumers and raises a great health concern worldwide.

Detection

Some *Salmonella* serovars are known to be pathogenic to humans, but all salmonellae are unacceptable in ready-to-eat foods. It is uncommon to find *Salmonella* in countable levels in food. Therefore, food is analyzed for the presence or absence of *Salmonella*, and detection methods rather than direct plating are used. Conventional methods for detection of *Salmonella* in food are based on the cultural, biochemical, and serological properties of the microorganism.

Alternative rapid methods are becoming popular in the food industry, and most of these are based on genetic or immunoassay techniques. These methods do not include the labor-intensive and time-consuming biochemical identification steps. The rapid methods still require sample enrichments, and only the negative results are considered conclusive. Positive results by rapid methods require verification using the conventional biochemical and serological identification techniques. In laboratories where a large number of samples are analyzed routinely, these alternative methods are useful in rapidly screening these samples for *Salmonella*. Since most commercial food samples are expected to be *Salmonella* free, use of rapid methods as a screening tool could save time and efforts.

Conventional Methods Conventional methods of detection of *Salmonella* in food depend on the cultural, biochemical, and serological properties of this microorganism. Typical *Salmonella* isolates are those that (i) produce acid from glucose, but not from lactose or sucrose, in triple sugar iron (TSI) agar medium; (ii) decarboxylate lysine to cadaverine (alkaline product) in lysine iron agar (LIA) medium; (iii) generate hydrogen sulfide (H_2S) in TSI agar and LIA; (iv) do not ferment lactose or sucrose in xylose lysine desoxycholate (XLD), Hektoen enteric (HE), and similar agar media; and (v) do not hydrolyze urea. Although *Salmonella*'s inability to ferment lactose or sucrose is an important defining biochemical property, some *Salmonella* isolates are lactose or sucrose positive. Fermentation of these sugars is coded on a plasmid that may be acquired or lost through conjugative genetic exchange.

Conventional methods involve (a) preenrichment and selective enrichment, (b) isolation steps, and (c) identification tests. Enrichment and isolation steps are primarily cultural techniques, whereas identification relies on biochemical and serological testing.

Enrichment Unlike clinical samples, foods contain only a small population of pathogens, if any. While isolation of *Salmonella* from clinical samples often requires only direct plating on selective agar media, detection of the pathogen in food necessitates an enrichment process to increase the numbers before isolation and identification can be accomplished. *Salmonella* in food is often present in an injured state. Cells are often injured during food processing (e.g., acidification and dehydration) and storage (e.g., freezing). Enrichment steps are normally designed to resuscitate these injured cells.

Preenrichment and selective enrichment steps are carried out to resuscitate injured salmonellae and increase their number to detectable levels. In the preenrichment step, the food sample is mixed with a suitable nonselective medium and the mixture is incubated. This results in a modest increase in *Salmonella* population and appreciable increases in numbers of competing microbiota in the sample. After the preenrichment, it is presumed that salmonellae are healthy enough to endure a subsequent enrichment step using media containing selective agents. Therefore, a small volume of the preenriched sample is transferred into the selective enrichment medium. This selective enrichment step allows growth of *Salmonella* and suppresses competing microbiota.

Media used in preenrichments vary with the type of food analyzed. Lactose broth is the most commonly used preenrichment medium, in spite of the fact that most *Salmonella* isolates do not use lactose as a carbon source. This medium, however, seems suitable for the slow recovery of injured cells in the food sample. Rich food such as dry milk may be enriched for *Salmonella* by simply mixing the sample with distilled water and incubating the mixture. In contrast, nutritionally poor foods (e.g., spices) are preenriched in *Salmonella* by adding a rich microbiological medium such as trypticase soy broth.

Selective enrichment includes subculturing the preenrichment into tubes containing selective broth. Selenite cystine broth and tetrathionate broth are widely used for selective enrichment. Composition and selective properties of these two media are reviewed later in this chapter. Rappaport–Vassiladis medium has been recommended as a replacement for the selenite cystine broth. It is advisable to use two or more selective enrichment media to improve the recovery of *Salmonella* from the sample.

Isolation Isolation includes streaking enrichments onto selective–differential agar media and recognizing presumptive *Salmonella* colonies on the incubated plates. These isolated colonies are subcultured for subsequent identification. Media for *Salmonella* isolation should contain selective agents such as bile or desoxycholate salts, brilliant green, and bismuth sulfite. These agents inhibit gram-positive and nonenteric bacteria. Differentiation of *Salmonella* and non-*Salmonella* bacteria is accomplished by inclusion of suitable carbohydrate/pH indicator combinations. Lactose, sucrose, and salicin are not typically fermented by *Salmonella*, and thus production of acid from these carbohydrates indicates non-*Salmonella* isolates. If lysine is present in the medium, *Salmonella* decarboxylates the amino acid, producing alkaline products that change the color of the pH indicator in the agar surrounding the colony. A familiar differential system in these media depends on the ability of *Salmonella* to release H_2S from sulfur-containing substrates (e.g., sodium thiosulfate) using its desulfhydrase. In addition to the enzyme substrate, the isolation media are formulated to contain water-soluble ferrous or ferric salt which reacts with the released H_2S to produce a black precipitate in and around the typical *Salmonella* colonies.

The principles just described were implemented in formulating several isolation media. Xylose lysine deoxycholate (XLD) agar and Hektoen enteric (HE) agar are commonly used for isolation of *Salmonella* from enrichments. Additionally, bismuth sulfite (BS) agar is an ideal medium for isolating salmonellae. It has high selectivity for the bacterium and allows detection of small levels of H_2S generated by isolated colonies. It is common to streak a *Salmonella*-enriched sample onto two or more of these selective–differential agar media; thus, recovery of pathogens from the food is improved and method sensitivity increases. Description and properties of these media are discussed later in this chapter.

Biochemical Identification   Conventional detection methods rely on biochemical tests to identify *Salmonella*. Decarboxylation of lysine in LIA, fermentation of glucose anaerobically in TSI agar, production of H_2S in TSI agar and LIA media, and utilization of citrates are important biochemical properties for identification of *Salmonella* isolates. Inability of *Salmonella* to (a) hydrolyze urea, (b) produce indole from tryptophan, and (c) grow in potassium cyanide broth are also useful biochemical tests for the characterization of the microorganism. Biochemical identification of *Salmonella*, therefore, requires testing presumptive isolates in TSI agar and LIA and running additional biochemical tests.

Serological Identification/Confirmation   Isolates that produce typical reactions by biochemical testing should be confirmed as *Salmonella* by serological analysis. If isolates react with antibodies developed against *Salmonella*'s somatic or flagellar antigens, this confirms the isolates as *Salmonella*. Isolates, therefore, are examined by a polyvalent flagellar test and a polyvalent somatic test; *Salmonella* produces agglutination upon testing. For serotyping, a more elaborate scheme of serological analysis is needed.

Rapid Methods   Rapid alternative detection methods for *Salmonella* require enrichment steps similar to those used in the conventional protocols. Additionally, a postenrichment step may be required to condition the isolates for the rapid technique. Time savings are achieved when these rapid methods are applied to screen a large number of samples and exclude the *Salmonella*-negative ones (Fig. 9.1). Positive results by the rapid methods still require confirmation by conventional techniques. Rapid methods for detection of *Salmonella* in food are based on immunological or genetic testing. Immunoassay-based and DNA hybridization-based methods are included in this chapter.

OBJECTIVES

- Detect *Salmonella* in food using a conventional method (the FDA protocol).
- Apply alternative rapid detection methods using molecular and serological techniques.
- Understand the merits and limitations of different detection methods.
- Practice safe handling of pathogens in the laboratory.

ORGANIZATION

Three laboratory exercises to detect *Salmonella* in food are presented in this chapter. During the first exercise, students analyze food using a conventional method, whereas rapid detection methods are used in the second and third exer-

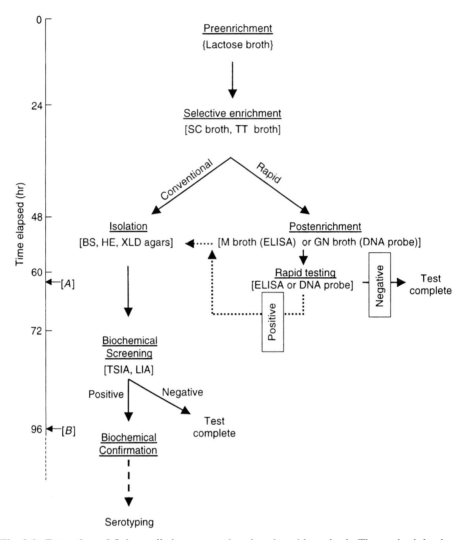

Fig. 9.1. Detection of *Salmonella* by conventional and rapid methods. Time schedule shown may be different than that followed in the laboratory exercises; it represents a minimum time frame. Labels of *y* axis: [A] time to complete a negative rapid test (~63 hr), [B] time to complete a negative conventional test (~96 hr).

cises (Table 9.1). The conventional method is completed in five laboratory periods while each of the rapid methods requires four periods. These three exercises may be run sequentially or simultaneously. Since the three methods share common enrichment steps, it saves time to run all three exercises side by side. In this case, however, provisions should be made not to run the ELISA and DNA probe steps in the same laboratory period because of the complexity of these techniques compared to the cultural protocol. Procedural details of each exercise are presented later in the chapter.

TABLE 9.1. Proposed Schedule[a] for Detection of *Salmonella* in Food Using Conventional (Culture–Biochemical–Serological) and Rapid Methods

Period	Conventional Method[b]	Immunoassay-Based Method[b]	Genetic-Based Method[b]
1	Preenrichment (in LB)	Preenrichment (in LB)	Preenrichment (in LB)
2	Selective enrichment (in TT and SC)	Selective enrichment (in TT and SC)	Selective enrichment (in TT and SC)
3	Isolation (on BS, HE, and XLD agar)	Postenrichment (in M broth)	Postenrichment (in GN broth)
4	Biochemical identification (on TSI agar and LIA slants)	ELISA test	DNA probe test
5	Inspection of TSI agar and LIA slants[c]		

[a] If the class does not meet daily, cultures that require only 24 h of incubation will be refrigerated until 24 h before the class meets and then incubated.
[b] If all three methods are executed simultaneously, ELISA and DNA probe tests may not be run during the same laboratory period because of the complexity of these techniques.
[c] Slants should be inspected after no longer than 24 h of incubation. Holding slants for additional incubation or in the refrigerator allows the spread of the black precipitation reaction, which makes it difficult to determine the result of the test.

PERSONAL SAFETY

All *Salmonella* serovars are considered pathogenic to humans and thus should be handled with care. Follow the safety guidelines that are reviewed at the beginning of this manual. Use disposable gloves when handling the isolates and the pathogenic cultures. Make sure the work area is sanitized after use.

REFERENCES

Andrews, W. H., and T. S. Hammack. 2000. *Bacteriological Analytical Mannual Online*, Chapter 5: *Salmonella*.
Available: http://www.cfsan.fda.gov/~ebam/bam_toc.html.

Andrews, W. H., R. S. Flowers, J. Silliker, and J. S. Baily. 2001. *Salmonella*. In F. P. Downes and K. Ito (Eds.), *Compendium of Methods for the Microbiological Examination of Foods*, 4th ed. (pp. 357–380). American Public Health Association, Washington, DC.

Centers for Disease Control and Prevention (CDC). July 2002. *Salmonella enteritidis*.
Available: http://www.cdc.gov/ncidod/dbmd/diseaseinfo/salment_g.htm.

D'Aoust, J.-Y., J. Maurer, and J. S. Baily. 2001. *Salmonella* Species. In M. P. Doyle, L. R. Beuchat, and T. J. Montville (Eds.), *Food Microbiology, Fundamentals and Frontiers*, 2nd ed. (pp. 141–178). American Society for Microbiology, Washington, DC.

Difco. 1998. *Difco Manual*, 11th ed. Difco Laboratories, Sparks, MD.

EXERCISE 1: CONVENTIONAL METHOD

PROCEDURE OVERVIEW

In this laboratory exercise, presence of *Salmonella* in food will be tested using a modification of the method described in the FDA *Bacteriological Analytical Manual* (BAM, 2000). The method depends on cultural, biochemical, and serological techniques and will be referred to as a "conventional/BAM" or simply "conventional" method. Two alternative rapid methods are also described later in this chapter as separate exercises. Since similar enrichment steps are applied in the three methods, it is advantageous to run two or all three methods simultaneously.

The conventional/BAM method includes five major steps: (i) preenrichment, (ii) selective enrichment, (iii) isolation on selective–differential agar media, (iv) identification and confirmation by biochemical testing using differential media, and (v) serological tests. In this laboratory exercise, the biochemical confirmation and serological testing will not be completed. An overview of this method, as applied in this laboratory exercise, is shown in Figs. 9.1 and 9.2. The detection by this conventional method will be completed in five laboratory periods, which correspond to four of the five steps just indicated (Table 9.1).

Results Interpretation

Positive results by the conventional detection method, as applied in this exercise, require that

1. *Salmonella* multiplies in the preenrichment (lactose broth) to a high enough level that some viable cells are transferred into the selective enrichment.
2. *Salmonella* is present in at least one of the selective enrichment tubes (SC or TT broth tubes) at a level suitable for transfer onto isolation agar media by streaking.
3. Enrichments result in a relatively high proportion of *Salmonella* to contaminants so that streaking on isolation agar plates (BS, HE, or XLD) produces at least one isolated colony with typical *Salmonella* characteristics on any of these plates.
4. One isolate at least, from the isolation agar plates, produces typical biochemical reactions of *Salmonella* on TSI agar and LIA.

MEDIA

Reactions of typical *Salmonella* isolates on these media are summarized in Table 9.2. Media composition is provided in Appendix C.

Bismuth Sulfite (BS) Agar

This is a selective–differential medium used for the isolation of potential *Salmonella* colonies. Brilliant green in the medium acts as a selective agent by inhibiting the growth of gram-positive organisms. Bismuth sulfite has both selective and differen-

CONVENTIONAL METHOD 175

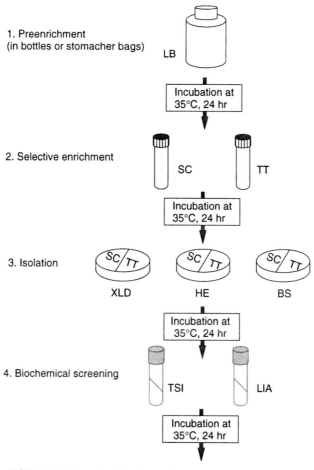

Fig. 9.2. Conventional method for detection of *Salmonella* in food.

tial effects. The bismuth selects against coliforms and also against gram positives. The sulfite allows differentiation of organism producing the enzyme desulfhydrase. Production of desulfhydrase by *Salmonella* results in formation of hydrogen sulfide (H_2S) from the sulfite. The H_2S formed then reacts with ferrous sulfate ($FeSO_4$) to form a black precipitate of ferrous sulfide (FeS). *Salmonella* colonies on BS agar appear black to green in color with or without a dark halo in the surrounding agar and the colonies often appear to have a metallic sheen.

Hektoen Enteric (HE) Agar

Hektoen enteric agar is a selective–differential medium used for the isolation of presumptive *Salmonella* colonies. Bile salts, bromothymol blue, and acid fuchsin are selective agents that inhibit gram-positive bacteria. Acid fuchsin and bromothymol blue also function as acid indicators detecting acid production from the fermenta-

TABLE 9.2. Characteristics of *Salmonella* on Isolation and Biochemical Identification Media

Medium and (Color)	*Salmonella* Reaction	*Salmonella* Interpretation	Other Microorganisms
Bismuth sulfite agar: light grey-green	Colonies black to green with or without dark halo and metallic sheen	*Salmonella* survives the strong selectivity of the medium; production of H_2S	Most are inhibited; *E. coli* may produce brown to green colonies
Hektoen enteric agar: green-yellowish	Colonies greenish blue with black center	No lactose, sucrose, or salicin fermentation; production of H_2S	Yellow to orange colonies (carbohydrate fermentation)
Xylose lysine desoxycholate agar: red	Red with black centers	Limited carbohydrate fermentation; lysine decarboxylation; production of H_2S	Yellow (carbohydrate fermentation); red with no black center (no H_2S production)
Triple sugar iron agar: red	Slant: red (alkaline)	No significant acid from carbohydrates; alkaline products from protein breakdown	Slant: yellow (acid)
	Butt: yellow (acid)	Glucose fermentation	Butt: red (alkaline)
	Black precipitate	Production of H_2S	No precipitate
Lysine iron agar: purple	Slant: purple (alkaline)	No acid from carbohydrates; alkaline products from protein breakdown	Slant: yellow (acid) or red (neutral)
	Butt: purple (alkaline)	Alkaline products from lysine decarboxylation	Butt: yellow (acid) or red (neutral)
	Black precipitate	Production of H_2S	No precipitate

tion of the carbohydrates (lactose, salicin, and sucrose) in the medium. Acid production by sugar fermentation changes the color of the pH indicator to yellow, and thus colonies become surrounded with yellow agar. *Salmonella* does not ferment these sugars but uses the proteose peptone as a source of energy with the production of alkaline end products. Colonies of *Salmonella*, therefore, appear greenish blue as the result of the indicator's color change. Ferric ammonium citrate and sodium thiosulfate combine to detect the production of desulfhydrase enzymes by *Salmonella*. In this case, the bacterial enzyme converts thiosulfate ($S_2O_3^{2-}$) to H_2S, which can then react with the ferric ion (Fe^{3+}) to create a black precipitate. *Salmonella* colonies on HE agar appear greenish blue with a black center.

Lactose Broth (LB)

Lactose broth is a nonselective medium. The lactose in the medium is not usable by *Salmonella*, but a lactose-containing medium seems to promote the slow recovery

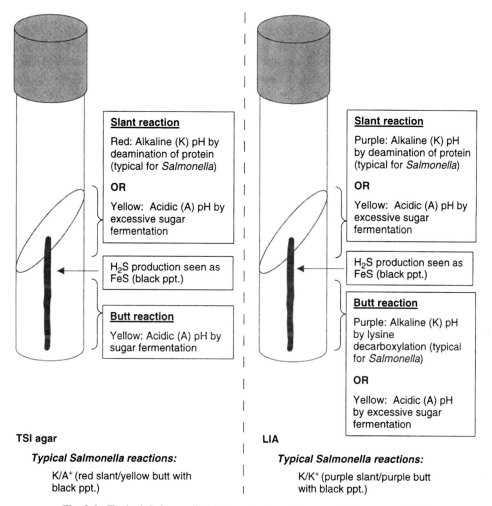

Fig. 9.3. Typical *Salmonella* biochemical reactions on TSI agar and LIA.

of injured *Salmonella* cells. Use of a broth medium dilutes the toxic or inhibitory substances in the food. The non-*Salmonella* populations increase considerably during incubation of the LB–food mixture. The numbers of salmonellae will also increase, but at a slower rate than that experienced by most other species present.

Lysine Iron Agar (LIA)

This is a differential agar medium that is dispensed in tubes to form a slant with a deep bottom (Fig. 9.3). Presumptive *Salmonella* isolates are streaked on the slant and stabbed in the tube butt for biochemical identification (Fig. 9.3). Reactions in the slant area are aerobic while those in the butt area are primarily anaerobic. The aerobic condition of the slant is assured by leaving the tube cap only lightly tightened.

The medium contains a limited amount of glucose, an ample supply of lysine, and a pH indicator (bromcresol purple). This pH indicator turns yellow in acidic environment, red at neutral conditions, and purple under alkaline conditions. *Salmonella* quickly utilizes the glucose first and turns the medium yellow because of acid production. After depletion of glucose, *Salmonella* utilizes lysine as an energy source. Salmonellae possess the enzyme lysine decarboxylase and a decarboxylation reaction occurs under anaerobic conditions (tube butt). The decarboxylation of lysine results in amine byproducts which produce alkaline reactions. The quantity of alkaline products is sufficient to neutralize the acid from glucose utilization and produce alkaline reaction in the tube butt (purple). In the aerobic slant, peptone metabolism is possible. When peptone is metabolized, this results in basic end products. These products turn the medium basic and thus a purple slant is observed. These fermentations may result is gas production in the LIA agar tube. If the gas bubbles are large enough, cracks can be formed, even breaking the agar into chunks. *Salmonella* may or may not generate enough gas to produce noticeable bubbles.

A differential system that consists of sodium thiosulfate and ferric ammonium citrate is included in this medium. Hydrogen sulfide results from the activity of *Salmonella*'s desulfhydrase on the thiosulfate. Released hydrogen sulfide reacts with the ferric ion, producing a black precipitate, particularly around the stab.

Selenite Cystine (SC) Broth

This medium is used for the selective enrichment of *Salmonella*. The selective agent, sodium acid selenite, is deleterious to most bacterial species. *Salmonella*, however, is more resistant to this chemical than are many other bacteria (such as fecal coliforms and enterococci). Because of this resistance, the medium is used to selectively enrich food samples in *Salmonella*. L-Cystine is added to the medium as a reducing agent.

Tetrathionate (TT) Broth

Like SC broth, this medium also is used for the selective enrichment of *Salmonella*. Iodine in the medium catalyzes the conversion of sodium thiosulfate to tetrathionate ($S_2O_3^{2-}$—$I_2 \rightarrow S_4O_6^{2-}$). Tetrathionate is toxic to many bacteria. Salmonellae are selected for because they possess an enzyme, tetrathionate reductase, that detoxifies the tetrathionate. Because salmonellae can detoxify the compound, they will multiply more rapidly in this medium than will other species. Bile salts in the medium also provide selection against nonenterics.

Triple Sugar Iron (TSI) Agar

This is a differential medium prepared as a slant agar in test tubes (Fig. 9.3). The presumptive *Salmonella* isolates are streaked on the surface of the slant and stabbed in the butt of tube contents. Reactions in the slant area are aerobic, and those in the butt are primarily anaerobic. When this medium is used, the tube cap fitting should be loosened. If screw-cap tubes are used, the caps must be tightened loosely to maintain aerobic conditions in the slant portion.

The medium contains three sugars, glucose, sucrose, and lactose, at 0.1, 1.0, and 1.0% levels, respectively. A pH indicator (phenol red) is included in the medium to detect acid production from fermentation of these carbohydrates. Phenol red changes medium color to yellow upon acid production by the inoculated microorganism while no color change indicates an alkaline surrounding. *Salmonella* metabolizes the limited supply of glucose (but not the sucrose or lactose) and produces acid in the relatively anaerobic butt of the TSI agar tube. *Salmonella* on the slant metabolizes glucose aerobically with no or little acid production. Additionally, aerobic breakdown of proteins by *Salmonella* results in basic end products that turn the slant alkaline. Microorganisms that metabolize lactose and sucrose produce strong acid reactions that turn the tube contents to yellow. Microorganisms may also produce carbon dioxide as a result of sugar metabolism. This results in bubbles in the agar. If the bubbles are large enough, cracks can be formed, even breaking the agar into chunks. *Salmonella* may or may not generate enough gas to produce noticeable bubbles.

The medium contains an additional differential system that consists of sodium thiosulfate ($Na_2S_2O_3$) and ferrous sulfate. *Salmonella*'s desulfhydrase enzyme converts the thiosulfate into hydrogen sulfide, which reacts with ferrous sulfate ($FeSO_4$), producing a black precipitate of ferrous sulfide (FS). This precipitate may "cloud" the entire butt region of the tube, making interpretation of butt reactions difficult, particularly when tubes are inadvertently incubated for longer than 24 hr.

Xylose Lysine Desoxycholate (XLD) Agar

This is a selective–differential medium used for the isolation of potential *Salmonella* colonies. The desoxycholate inhibits nonenterics. This medium uses ferric ammonium citrate and sodium thiosulfate as differential agents in the same way as does HE agar, producing the same possible result. A pH indicator (phenol red) is used to detect the fermentation of sugars (xylose, lactose, and sucrose) in the medium. Fermentation, and therefore acid production by these carbohydrates, turns this medium yellow. Of these sugars, *Salmonella* ferments only xylose. L-Lysine is added to the medium to differentiate salmonellae from other xylose fermenters. Salmonellae have the ability to decarboxylate lysine. Because *Salmonella* ferments a limited amount of sugar (only the xylose), the decarboxylation reaction results in sufficient basic products to balance and overwhelm the acidic products of xylose fermentation. This alkaline reversion turns the colonies and medium around them red. Therefore, the combination of pH indicator, sugars, and amino acid allows differentiation of organisms by metabolic activity. *Salmonella* colonies on XLD agar, therefore, are red with black centers.

ORGANIZATION

Each pair of students will test one food sample. Retail packages of ground meat (chicken, turkey, beef, or pork) will be tested for *Salmonella* using the conventional/BAM protocol. Two rapid methods, enzyme-linked immunosorbent assay (ELISA) and DNA-based probe, may also be run concomitantly with the conventional/BAM method.

Period 1 Preenrichment

MATERIALS AND EQUIPMENT

Per Pair of Students

- Food sample
- Culture of *Salmonella* (positive control)
- Lactose broth
  a. One 225-ml flask for food enrichment
  b. One 10-ml tube for *Salmonella* enrichment (control)

Class Shared

- Scale for weighing food samples (e.g., a top-loading balance with 500 g capacity)
- Incubator, set at 35°C
- Stomacher and stomacher bags
- Container for holding enrichment bags

PROCEDURE

Control

1. Transfer 1 ml of the *Salmonella* culture to the 10-ml lactose broth tube.
2. Incubate the inoculated lactose broth tube at 35°C for 24 hr. The incubated mixture in these tubes will be referred to as "control–LB enrichment."

Food Sample

1. Prepare the food for sampling to ensure that a representative sample is analyzed. This varies with the food and may involve cutting, partitioning, grinding, or mixing.
2. Weigh 25 g of the food sample into a stomacher bag. Add the 225 ml lactose broth. Stomach for 2 min.
3. Place the enrichment-containing bag in the tub or beaker designated by the laboratory instructor. This container should support the bag and keep it upright during incubation and handling.
4. Incubate the bags at 35°C for 24 hr. The incubated mixture in these bags will be referred to as "food–LB enrichment."

Period 2 Selective Enrichment

MATERIALS AND EQUIPMENT

Per Pair of Students

- Control–LB enrichment
- Food–LB enrichment
- Two 10-ml tubes selenite cystine broth
- Two 10-ml tubes tetrathionate broth

Class Shared

- Incubator, set at 35°C

PROCEDURE

Positive Control

1. Label one selenite cystine and one tetrathionate broth tube for the positive control.
2. Mix the content of the control–LB enrichment tube and transfer 1 ml into each of the selenite cystine and tetrathionate broth tubes.
3. Incubate the inoculated tubes at 35°C for 24 hr. The contents of the incubated tubes will be referred to as "control–SC enrichment" and "control–TT enrichment," respectively.

Food Sample

1. Label one selenite cystine and one tetrathionate broth tube for the food sample.
2. Mix the food–LB enrichment by hand agitating the stomacher bag.
3. Transfer 1 ml food–LB enrichment into the selenite cystine broth tube; repeat with the tetrathionate broth tube.
4. Incubate the inoculated tubes at 35°C for 24 hr. The contents of the incubated tubes will be referred to as "food–SC enrichment" and "food–TT enrichment," respectively.

Period 3 Isolation on Selective Agar Media

MATERIALS AND EQUIPMENT

Per Pair of Students

- Control–SC and control–TT enrichment tubes
- Food–SC and food–TT enrichment tubes
- Bismuth sulfite agar (two plates)
- Hektoen enteric agar (two plates)
- Xylose lysine desoxycholate agar (two plates)

Class Shared

- Incubator, set at 35°C

PROCEDURE

1. Label one BS plate for the food sample and the other for the positive control. Divide each plate into halves, one for SC enrichment as the source and the other for the TT enrichment as the source. Repeat for the HE and XLD plates.
2. Two-phase streak the food–SC enrichment onto the appropriate half of the BS, HE, and XLD plates. Two-phase streak the food–TT enrichment onto the other half of the BS, HE, and XLD plates.
3. Repeat step 2 using the positive control (control–SC and control–TT enrichments).
4. Incubate the streaked plates at 35°C for 24 hr.

Period 4 Biochemical Identification

MATERIALS AND EQUIPMENT

Per Pair of Students

- BS, HE, and XLD incubated plates (prepared during the previous laboratory period)
- Lysine iron agar slants (seven tubes)
- Triple sugar iron agar slants (seven tubes)

Class Shared

- Incubator, set at 35°C

PROCEDURE

Inspection of BS, HE, and XLD Plates

BS Agar
Positive Control

1. Observe the BS plate inoculated with the positive control. Typical *Salmonella* colonies are black to green with or without a dark halo and metallic sheen.
2. Compare the appearance of colonies originating from the SC and TT enrichments. Note any differences in appearance or prevalence.
3. Record the observations in Table 9.3.

Food Sample

1. Observe the BS plate inoculated with the enriched food sample. Compare the appearance of these colonies with that of the positive control. Mark the location of possible isolated colonies of *Salmonella*.
2. Compare the appearance of colonies originating from the SC and TT enrichments. Note any differences in appearance or prevalence.
3. Record the observations in Table 9.3.

HE Agar
Positive Control

1. Observe the HE plate inoculated with the positive control. Typical *Salmonella* colonies are greenish blue with black center.
2. Compare the appearance of colonies originating from the SC and TT enrichments. Note any differences in appearance or prevalence.
3. Record the observations in Table 9.3.

Food Sample

1. Observe the HE plate inoculated with the enriched food sample. Compare the appearance of these colonies with that of the positive control. Mark location of possible isolated colonies of *Salmonella*.

TABLE 9.3. *(Write a descriptive title, including source of inoculum, previous manipulations, and results presented)*

Plating Medium	Enrichment Medium	Appearance	
		Food Sample	Positive Control
BS	SC		
	TT		
HE	SC		
	TT		
XLD	SC		
	TT		

(Add at least five footnotes).

2. Compare the appearance of colonies originating from the SC and TT enrichments. Note any differences in appearance or quantity.
3. Record the observations in Table 9.3.

XLD Agar
Positive Control

1. Observe the XLD plate inoculated with the positive control. Typical *Salmonella* colonies are red-pink colonies. These colonies commonly have a black center.
2. Compare the appearance of colonies originating from the SC and TT enrichments. Note any differences in appearance or quantity.
3. Record the observations in Table 9.3.

Food Sample

1. Observe the XLD plate inoculated with the enriched food sample. Compare the appearance of these colonies with that of the positive control. Mark the location of possible isolated colonies of *Salmonella*.
2. Compare the appearance of colonies originating from the SC and TT enrichments. Note any differences in appearance or quantity.
3. Record the observations in Table 9.3.

Biochemical Identification

1. Label six tubes of each slant medium (TSI agar and LIA) for use with the enriched food sample. The label should include the medium from which the colony was obtained (BS, HE, XLD) and the type of enrichment (SC, TT).

Label one tube of TSI agar and one of LIA for the positive control. There will be a total of seven tubes of each medium.
2. Choose and mark two possible *Salmonella* colonies, if any are present, from each of the plates (BS, HE, and XLD) that were inoculated from the enriched food sample. One should be from the SC half and the other from the TT half. This is a total of six colonies.
3. For each chosen colony, inoculate both a TSI agar and an LIA slant.
    a. Using a sterile inoculating needle, lightly touch the center of the selected colony.
    b. Inoculate the agar slant by beginning at the base of the slant and working upward in a zigzag pattern. Do not streak higher than two-thirds of the way up.
    c. Next, without flaming the loop, stab the needle into the butt, stopping 1–2 cm from the base of the tube.
4. Inoculate one TSI agar and one LIA slant from any representative colony of the positive control.
5. Incubate the inoculated slants at 35°C for 24 hr.

186 *Salmonella*

Period 5 Biochemical Identification (*Continued*)

MATERIALS AND EQUIPMENT

- Inoculated and incubated TSI agar and LIA tubes
- Uninoculated tube of TSI agar and another of LIA (negative controls)

PROCEDURE

Biochemical Testing Results

Observe the tubes. It is recommended that tubes are placed in a rack to make comparisons easier. For each tube, determine the slant and butt reactions and determine if a black precipitate has formed (Fig. 9.3). Report these results in Table 9.4.

Interpretation Guide

TSI Agar: Salmonella can use glucose and will use peptone aerobically, and most serovars produce desulfhydrase enzymes. A *Salmonella*-positive result is an alkaline (red) slant and an acid (yellow) butt, probably with black precipitate present. This is reported as K (alkaline) for the slant and A (acid) for the butt, with the production of precipitate indicated by a superscript plus or minus, as applicable. Sometimes gas may also be produced. Report this by including a "g" as a superscript. *Examples*: K/A$^-$ or K/A^{+g}. Figure 9.3 shows an interpretation of reaction on TSI agar.

TABLE 9.4. (*Write a suitable title, including food, meaning of plate/enrichment medium, and meaning of results*)

Plating Medium	Enrichment Source	TSI Agar	LIA
BS	SC		
	TT		
HE	SC		
	TT		
XLD	SC		
	TT		
Salmonella control	SC or TT		

Key: (K) alkaline, (A) acid, (+) precipitate found, (g) gas formed.
(*Add appropriate footnotes*).

LIA: *Salmonella* uses glucose, will use peptone aerobically, and possesses the enzyme lysine decarboxylase, and most serovars produce desulfhydrase enzymes. A *Salmonella*-positive result would be a purple (alkaline) slant and a purple (alkaline) butt, probably with black precipitate present. This is reported as K (alkaline) for the slant and K for the butt, with the production of precipitate indicated by a superscript plus or minus, as applicable. Sometimes gas may also be produced. Report this by including a "g" as a superscript. Neutral (red) reactions can be reported as "N." *Example*: K/K$^+$ (Fig. 9.3).

Determine whether the food sample is presumptively positive for the presence of *Salmonella*; this requires at least one tube showing typical *Salmonella* reactions. Record these data on a form provided by the laboratory instructor.

EXERCISE 2: IMMUNOASSAY-BASED METHOD

PROCEDURE OVERVIEW

A rapid method with an ELISA test will be used in this exercise to detect *Salmonella* in food. The method includes preenrichment and selective enrichment steps similar to those applied in the conventional detection protocol (Table 9.1, Fig. 9.4). An additional enrichment (a postenrichment) in mannose broth precedes the immunoassay step. Using the ELISA-based detection method to cull out *Salmonella*-free food samples saves considerable amount of time (Fig. 9.1).

A commercial ELISA test kit (*Salmonella*-Tek, bioMérieux, St. Louis, MO) will be used for rapid screening of *Salmonella* in food samples. This kit includes polystyrene microtiter strips with wells precoated with antibodies against *Salmonella* antigens. According to the kit manufacturer, the microtiter plate wells are coated with monoclonal antibodies suitable for detection of both flagellated and nonflagellated *Salmonella* serovars. The kit also includes a conjugate, an enzyme-labeled secondary antibody. The antibody portion allows the conjugate to bind *Salmonella* antigen present in the well and to be retained after washing. If no *Salmonella* antigens are present in the food sample, the conjugate is washed out during the washing step. The enzyme portion of the conjugate (horseradish peroxidase, HRP) catalyzes the reaction between hydrogen peroxide and 3,3′,5,5′-tetramethylbenzidine (TMB). This reaction results in a blue-colored product. Sulfuric acid is added to the mixture to stop the reaction; this changes the product color from blue to yellow:

$$H_2O_2 + TMB \xrightarrow{(HRP)} \text{Blue product} \xrightarrow{(\text{Stop solution})} \text{Yellow}$$

Comparing the amount of color in the control wells to the amount of color in the sample wells (using a microtiter plate reader) allows determination as to whether the sample is negative or positive for *Salmonella*.

Kit-negative is a solution containing a non-*Salmonella* antigen. Kit-positive is a solution containing *Salmonella* antigen. Use of the Kit-positive and Kit-negative samples allows the user to determine if the wells were adequately washed (i.e., Kit-negative has negative result) and the reagents are working properly (i.e., Kit-positive produces a positive result).

The immunoassay-based method rapidly screens food samples that are *Salmonella* negative (Fig. 9.1), and such a method is described as a "negative" test. Its primary use is to eliminate negative samples more rapidly than when using the conventional cultural method. Positive results should be analyzed further using the conventional cultural protocol. Notice also that *Salmonella*-Tek ELISA cannot determine the serovar of *Salmonella*. An overview of the procedure is shown in Fig. 9.4. Before running this exercise, analysts should check the commercial kit interest for complete instructions regarding reagent preparation and for the latest procedural modifications.

MEDIA

Mannose (M) Broth

This medium enriches for *Salmonella*, although it is not as selective as SC or TT broths. Mannose is a carbohydrate that is usable by salmonellae. Sodium citrate

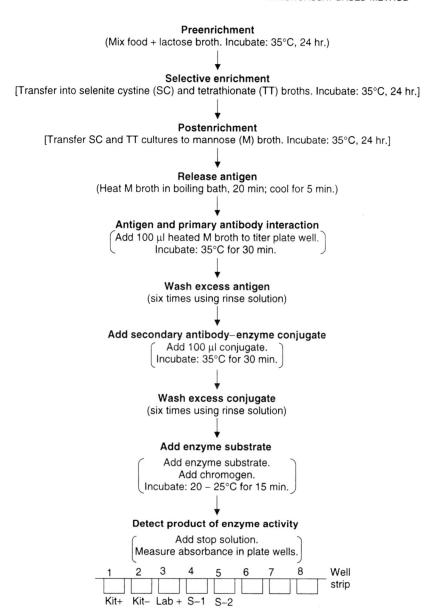

Fig. 9.4. Immunoassay-based method for detection of *Salmonella* in food.

weakly selects against nonenterics. The medium is also differential as it contains both magnesium sulfate and ferrous sulfate. The desulfhydrase reactions described in the plating media may also occur in M broth. Therefore, tubes with a blackish precipitate are likely to contain salmonellae. The medium is required to yield growth of *Salmonella* rapidly and efficiently and to prevent nonspecific agglutination. Supplementing M broth with 10 µg/ml novobiocin improves recovery of *Salmonella* from raw foods by inhibiting the growth of competitive organisms.

ORGANIZATION

Retail packages of ground meat (chicken, turkey, beef, or pork) will be tested for *Salmonella* using the ELISA-based method (Figs. 9.1 and 9.4). Each pair of students will carry out the enrichment steps using a food sample. Subsequently, four students, with two food samples, will run the ELISA test and share an eight-well strip. Notice that the kit microtiter plates come with removable rows of wells (strips).

Period 1 Preenrichment

Use the same steps as in the conventional method.

Period 2 Selective Enrichment

Use the same steps as in the conventional method.

Period 3 Postenrichment

In this laboratory period, a third enrichment step (postenrichment) will be completed. Postenrichment involves subculturing the selectively enriched samples from the incubated SC and TT broths to M broth.

MATERIALS AND EQUIPMENT

Per Pair of Students

- Control–SC and control-TT enrichment tubes
- Food–SC and food–TT enrichment tubes
- Four 10-ml tubes M broth

Class Shared

- Incubator, set at 35°C

PROCEDURE

Positive Control

1. Label two M-broth tubes, one receiving the control–SC enrichment (control–SC–M) and the other the control–TT enrichment (control–TT–M).
2. Mix the contents of the control–SC enrichment tube. Mix also the contents of the control–TT enrichment tube.
3. Transfer 1 ml control–SC enrichment into the control–SC–M broth tube.
4. Similarly, transfer 1 ml control–TT enrichment into the control–TT–M broth tube.
5. Incubate the inoculated tubes at 35°C for 24 hr.

Food Sample

1. Label two M-broth tubes, one receiving the food–SC enrichment (food–SC–M) and the other the food–TT enrichment (food–TT–M).
2. Mix the contents of the food–SC enrichment tube. Mix also the contents of the food–TT enrichment tube.
3. Transfer 1 ml food–SC enrichment into the food–SC–M broth tube.
4. Similarly, transfer 1 ml food–TT enrichment into the food–TT–M broth tube.
5. Incubate the inoculated tubes at 35°C for 24 hr.

Period 4 Enzyme-Linked Immunosorbent Assay

MATERIALS AND EQUIPMENT

Complete instructions for preparing reagents used in this laboratory exercise and amounts of each needed to run the test can be found in the package insert provided by the commercial kit producer.

Per Group of Four Students

- Enrichments for two food samples: a food–SC–M and a food–TT–M enrichment tube for each of the two food samples (prepared during the previous laboratory period)
- Positive control enrichments: control–SC–M and control–TT–M enrichment tubes (prepared during the previous laboratory period)
- One microcentrifuge tube containing Kit-positive
- One microcentrifuge tube containing Kit-negative
- Three sterile screw-cap microcentrifuge tubes
- Wash solution, 40 ml
- One microcentrifuge tube containing conjugate
- One microcentrifuge tube containing TMB-A (3,3',5,5'-tetramethylbenzidine)
- One microcentrifuge tube containing TMB-B (hydrogen peroxide)
- One microcentrifuge tube containing stop solution
- ELISA test strip of eight wells
- Microtiter plate sealer strips
- Pipetter (0–1000 µl capacity) and tips

Class Shared

- Incubator, set at 35°C
- Microtiter plate reader

PROCEDURE

Sample Preparation

1. *Food-1*: Mix the contents of the food–SC–M enrichment tube. Mix also the contents of the food–TT–M enrichment tube. Combine 0.5 ml of food–SC–M enrichment and 0.5 ml of food–TT–M enrichment in a screw-cap microcentrifuge tube. Label this tube as "Food-1."
2. *Food-2*: Repeat step 1 using enrichment tubes for the second food sample. Label the resulting microcentrifuge tube as "Food-2."
3. For each group of four students, prepare one *Salmonella*-positive control using 0.5 ml from the control-SC-M and 0.5 ml from the control–TT–M enrichment tubes. Combine these portions in a microcentrifuge tube and label as "Lab-positive."

4. Heat Food-1, Food-2, and Lab-positive tubes in a boiling water bath for 20 min.
5. Cool the heated tubes to room temperature (for approximately 5 min).

Well Preparation

The following well labeling order of the eight-well strip may be followed:

#1: Kit-positive
#2: Kit-negative
#3: Lab-positive
#4: Food-1
#5: Food-2
#6, 7, 8: Empty

ELISA

1. Pipette 100 µl of Food-1, Food-2, or controls into the appropriate well (see the order suggested earlier).
2. Cover the well strip with a plate sealer.
3. Incubate the well strip at 35°C for 30 min.
4. Wash each well used six times with wash solution; use 100 µl per well for each wash. Do not overfill as this can result in wells contaminating each other. If the pipette tip touches a well, change the tip. Try to empty the wells completely before refilling. Do not let wells dry out. Unused wells do not require washing.
5. Pipette 100 µl of conjugate into each well being used. If the pipette tip touches a well, change tips.
6. Cover the plate with a plate sealer.
7. Incubate the plate at 35°C for 30 min.
8. Wash each well used six times with wash solution as indicated earlier (step 4).
9. Mix the TMB substrate by combining the contents of the TMB-A and TMB-B tubes. This mixing should be done only when ready to use the substrate.
10. Pipette 100 µl of TMB substrate into each well being used. If the pipette tip touches a well, change tips.
11. Incubate at room temperature for 15 min.
12. Pipette 100 µl stop solution into each well being used.
13. Read wells 5 min after adding stop solution.

Reading Strips

1. Read the well strip in a microtiter plate reader.
2. Record the results of the ELISA test in Table 9.5.

TABLE 9.5. *(Include a suitable title)*

Well Content	Absorbance (450 nm)	Valid?	Presumptive Result
Kit-positive			NA
Kit-negative			NA
Lab-positive			
Food-1			
Food-2			

(Add multiple footnotes as required for this table).

Interpreting Data

1. Calculation of the cutoff value: According to the kit manufacturer, ELISA plate readings are valid if the following conditions are met:
    a. The negative control (Kit-negative) must have an absorbance of less than 0.300.
    b. The positive control (Kit-positive) absorbance value must be at least 0.700.
    c. Values for Kit-positive and Kit-negative that fall outside the manufacturer's predetermined limits indicate errors in protocol (such as underwashing wells) or problems with the reagents.

The cutoff value is calculated by adding 0.250 to the absorbance value of the Kit-negative control. Using this calculated cutoff value results in fewer false-positive results.

2. Use of the cutoff value:
    a. A sample with an absorbance value greater than or equal to the cutoff value is considered to be a presumptive positive result.
    b. A sample with an absorbance reading below the cutoff value is considered to be a negative result.

EXERCISE 3: GENETIC-BASED METHOD

PROCEDURE OVERVIEW

A DNA probe-based method will be used in this exercise to rapidly detect *Salmonella* in food. If the food sample is *Salmonella* free, using the DNA probe-based detection method saves considerable assay time (Fig. 9.1). A commercial DNA probe kit (Gene-Trak *Salmonella* Assay, Neogen, Lansing, MI) will be used in this exercise. The kit includes a probe made of oligonucleotides that hybridize with sequences unique to *Salmonella*. A 16S rRNA sequence is used as the target because rRNA occurs in large copy number in cells. Additionally, when conserved sequences in rRNA are targeted, this increases test specificity.

The method includes preenrichment and selective enrichment steps similar to those applied in the conventional detection protocol (Table 9.1, Fig. 9.5). This is followed by an additional enrichment in Gram-negative (GN) broth (postenrichment). Cells in the incubated GN broth are lysed to release rRNA that subsequently binds to the probe (Fig. 9.6). Two probes are used; each targets a specific sequence on *Salmonella* rRNA. A capture probe hybridizes with the bacterial rRNA and binds it to a solid support (a plastic dip stick). This probe has a long poly-A tail that will bind to the poly-T coated dipstick, allowing the transfer of *Salmonella* rRNA from medium to the stick. The other probe (detector probe) is part of the detector system and is labeled with fluorescein. The detector probe hybridizes with the corresponding sequence on *Salmonella* rRNA. When the dipstick is transferred to a medium containing the enzyme conjugate (antifluorescein antibody, labeled with horseradish peroxidase), the conjugate binds the detector probe. Subsequently, the dipstick is transferred to the tube containing the substrates for horseradish peroxidase (hydrogen peroxide and TMB), and the enzyme catalyzes the reaction between these substrates, producing a blue-colored product. The amount of blue color developed is proportional to the amount of enzyme conjugate bound to the complex on the dipstick and also is proportional to the amount of targeted rRNA. The reaction is stopped by addition of sulfuric acid, which changes the blue color to yellow. Measuring the intensity of the yellow color at 450 nm allows detection of targeted rRNA and consequently the detection of *Salmonella* in the sample. Figure 9.6 shows how the binding of the components of the DNA probe results in a detectable reaction.

Analysts should prepare reagents according to instruction provided by the commercial test kit manufacturer. It is common that manufacturers of these kits make procedural changes or substitute reagents. Therefore, it is important that analysts check kit documentation for complete instructions and modifications.

MEDIA

Gram-negative (GN) Broth

This broth selectively enriches for *Salmonella* and against gram-positive bacteria as the medium contains sodium desoxycholate. Presence of mannitol in the medium

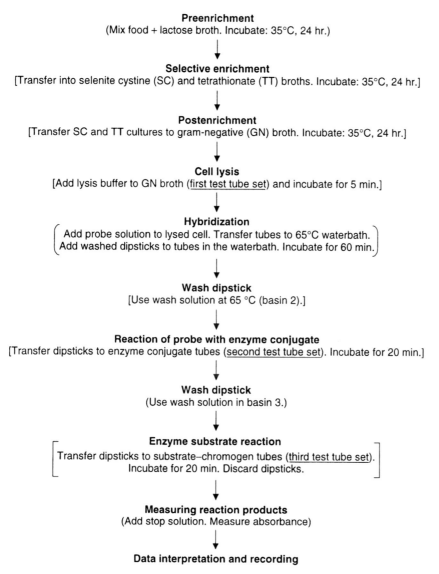

Fig. 9.5. The DNA probe-based method for detection of *Salmonella* in food.

selectively enriches *Salmonella* in the sample since the microorganism ferments this carbohydrate.

ORGANIZATION

Retail packages of ground meat (chicken, turkey, beef, or pork) will be tested for *Salmonella* using a rapid method that employs a DNA probe technique (Table 9.1 and Fig. 9.5). Each pair of students will carry out the enrichment steps using a food sample. Subsequently, four students, with two food samples, will run the DNA probe test and share a stick holder, which carries up to five dipsticks.

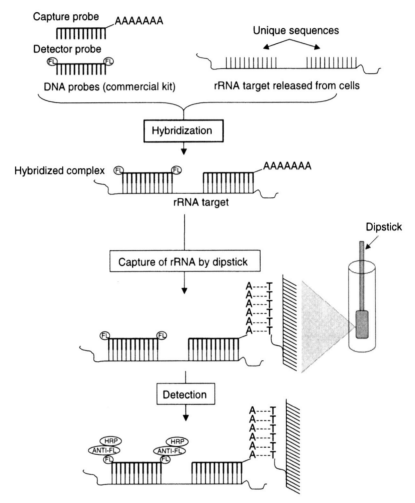

Fig. 9.6. DNA hybridization using the Gene-Trak *Salmonella* assay kit. (Adapted from Neogen Corp. commercial kit insert.)

Period 1 Preenrichment

Use the same steps as in the conventional method.

Period 2 Selective Enrichment

Use the same steps as in the conventional method.

Period 3 Postenrichment

In addition to the two enrichment steps, which are common among the conventional and rapid methods, a third enrichment step (postenrichment) will be completed. Postenrichment involves subculturing the selectively enriched samples from the incubated SC and TT broths to the GN broth.

MATERIALS AND EQUIPMENT

Per Pair of Students

- Food–SC and food–TT enrichment tubes
- Control–SC and control–TT enrichment tubes
- Four 10-ml tubes GN broth

Class Shared

- Incubator, set at 35°C

PROCEDURE

Food Sample

1. Label two GN broth tubes; one will receive the food–SC enrichment (food–SC–GN) and the other the food–TT enrichment (food–TT–GN).
2. Mix the contents of the food–SC enrichment tube. Mix also the contents of the food–TT enrichment tube.
3. Using a 1-ml inoculum, inoculate the food–SC–GN broth tube from the food–SC enrichment tube.
4. Using a 1-ml inoculum, inoculate the food–TT–GN broth tube from the food–TT enrichment tube.
5. Incubate the inoculated tubes at 35°C for 24 hr.

Positive Control

1. Label two GN broth tubes, one will receive the control–SC enrichment (control–SC–GN) and the other the control–TT enrichment (control–TT–GN).
2. Mix the contents of the control–SC enrichment tube. Similarly, mix the contents of the control–TT enrichment tube.
3. Using a 1-ml inoculum, inoculate the control–SC–GN broth tube from the control–SC enrichment tube.
4. Using a 1-ml inoculum, inoculate the control–TT–GN broth tube from the control–TT enrichment tube.
5. Incubate the inoculated tubes at 35°C for 24 hr.

Period 4 DNA Probe Hybridization

(*Note*: This method requires approximately 3 hr to complete.)

MATERIALS AND EQUIPMENT

Per Group of Four Students

- Enrichment tubes for two food samples: a food–SC–GN and a food–TT–GN enrichment tube for each of the two food samples (prepared during the previous laboratory period)
- Positive control enrichment tubes: control–SC–GN and control–TT–GN enrichment tubes (prepared during the previous laboratory period)
- Sixteen test tubes (12 × 75 mm)
- Five dipsticks (test kit content)
- One microcentrifuge tube containing Kit-positive
- One microcentrifuge tube containing Kit-negative
- One microcentrifuge tube containing lysis reagent
- One microcentrifuge tube containing *Salmonella* probe
- One microcentrifuge tube containing enzyme conjugate (*Caution*: This should be prepared immediately before use, preferably by laboratory instructors.)
- One microcentrifuge tube containing substrate chromogen
- One microcentrifuge tube containing stop solution
- Pipetter and tips (0–1000 µl capacity)

Class Shared

- Spectrophotometer (a dedicated spectrophotometer from Gene-Trak may be used)
- Waterbath set at 65°C

PROCEDURE

Preparation of Dipsticks and Reagents

1. Three basins of wash solutions are prepared according to the instructions in the kit insert, preferably by laboratory instructors before the laboratory period starts. Two of these basins will be kept at room temperature (basins 1 and 3) and the other in the 65°C waterbath (basin 2).
2. Prepare lysis solution as instructed in the kit insert. Alternatively, laboratory instructors may prepare the lysis solution before the laboratory period starts to save time.
3. Obtain a dipstick holder. Label the five slots on the holder as follows: Food-1, Food-2, Lab-positive, Kit-positive, and Kit-negative.

4. Obtain five dipsticks and insert on the dipstick holder. Handle the dipstick only by the flat handle part (never by the fins).
5. Place the dipstick assembly in the basin containing wash solution (basin 1) at room temperature.

Preparation of Sample and Controls; Cell Lysis

1. Prepare a rack of five test tubes (first test tube set) and label the tubes to correspond to the dipsticks on the holder (i.e., Food-1, Food-2, Lab-positive, Kit-positive, and Kit-negative).
2. Mix the contents of the food–SC–GN enrichment tube. Similarly, mix the contents of the food–TT–GN enrichment tube.
3. Pipette 0.25 ml from the food–SC–GN enrichment and 0.25 ml from the food–TT–GN enrichment and combine in a test tube labeled Food-1. Mix the contents of the tube.
4. Repeat steps 2 and 3 for the second food sample. Label the resulting sample tube as Food-2.
5. Repeat steps 2 and 3 for the *Salmonella*-positive control using the control–SC–GN and control–TT–GN enrichments. Label the resulting control tube as Lab-positive.
6. Dispense 0.5 ml of the kit's positive control and 0.5 ml of the kit's negative control to the corresponding test tubes. (*Note*: Because of the limited amount of these reagents in the commercial kit, it is preferable that laboratory instructors dispense these solutions in tubes before the start of the laboratory period. A student group may receive one or both controls based on the availability of these reagents.)
7. Add 0.1 ml lysis buffer to each of the five tubes.
8. Mix each tube gently by hand rolling.
9. Incubate for 5 min at room temperature.

Hybridization

1. Add 0.2 ml of *Salmonella* probe solution to each of the five tubes (containing the lysed samples and controls). Mix each tube gently by hand rolling.
2. Place the rack of tubes immediately in the 65°C waterbath. Be sure the tubes are arranged in the order matching the labeling on the dipsticks.
3. Remove the dipstick holder (and dipsticks) from the wash basin. Blot the tips lightly on a paper towel.
4. Insert the dipsticks in the tubes held in the 65°C waterbath. The holder should allow them to sit in the tube without touching the bottom. Move the dipstick up and down five times to mix.
5. Incubate in the 65°C waterbath for 60 min.

Reaction of Probe with Enzyme Conjugate

1. Label a second set of test tubes (a total of five) to match the slot labels on the dipstick holder.

2. Add 0.75 ml enzyme conjugate to each test tube. Remember this is mixed and becomes available just before it is needed.
3. Remove the dipsticks from the first set of test tubes after the 60 min incubation is complete. Discard properly the first test tube set.
4. Wash the dipsticks in the 65°C wash solution (basin 2). Blot the dipsticks on a paper towel.
5. Place the dipsticks in the second set of test tubes.
6. Incubate at room temperature for 20 min.

Enzyme–Substrate Reaction

1. Label six test tubes—five to match the slot labels on the dipstick holder and one for the reagent blank (this is the third set of tubes needed for this test).
2. Add 0.75 ml substrate–chromogen into each test tube.
3. Remove the dipsticks from the second set of test tubes, after the 20-min incubation is complete. Dispose of the second set of tubes.
4. Wash the dipsticks by moving them gently back and forth in the second room temperature wash solution basin (basin 3) for 1 min. Blot the dipsticks on a paper towel.
5. Place the dipsticks in the third set of test tubes. The reagent blank will not have a dipstick.
6. Incubate at room temperature for 20 min. Notice the development of a blue color in tubes with potentially positive samples.

Measuring Reaction Products

1. Remove the dipsticks from the test tubes. *Discard the dipsticks.*
2. Add 0.25 ml stop solution to each test tube.
3. Measure absorbance (at 450 nm) of the six tubes in a double-beam spectrophotometer. The following measurements are taken using the reference and measuring compartments, respectively:
   a. Reagent blank vs. Kit-negative
   b. Reagent blank vs. Kit-positive
   c. Kit-negative vs. Food-1
   d. Kit-negative vs. Food-2
   e. Kit-negative vs. Lab-positive

Results Interpretation

1. Negative control (Kit-negative) must have an absorbance value of less than or equal to 0.15 when read against the reagent blank in order for the results to be valid.
2. Positive control (Kit-positive) must have an absorbance value greater than or equal to 1.00 when read against the reagent blank.

TABLE 9.6. *(Include a suitable title)*

Tube Content	Absorbance (450 nm)	Valid?	Presumptive Result
Kit-negative			
Kit-positive			
Lab-positive			
Food-1			
Food-2			
Lab-positive			

(Add footnotes as required for this table).

3. If the sample (food or lab-positive) has an absorbance value less than or equal to 0.1, when read against a valid negative control, the result is interpreted as negative for *Salmonella*.
4. If the sample absorbance value is greater than 0.1, when read against a valid negative control, then the result is presumptive positive for *Salmonella*. Additional confirmation of this result using the conventional protocol is required.

Data Recording

Record the results of the DNA probe test in Table 9.6.

PROBLEMS

1. Using the data gathered in the previous three laboratory exercises, would it be possible to obtain quantitative information about the level of *Salmonella* contamination in the food? Explain.

2. List the biochemical characteristics of *Salmonella* that are examined in these exercises. Indicate the test or medium used in these laboratory exercises to check each of these properties.

3. For the food tested in this laboratory by you or your group, list the following information:
   (a) Method of packaging
   (b) Storage method
   (c) Sell-by date

4. Using the conventional method:
   (a) Fill in Table 9.3 with food sample and control results from BS, XLD, and HE plates. Add a suitable table title and footnotes.
   (b) Fill in Table 9.4 with food sample and control results from LIA and TSI agar tubes (slant/butt reaction–H$_2$S result format). Add a suitable table title and footnotes.
   (c) Determine and record the presumptive result for the food tested.
   (d) No negative control was used when food was tested with the conventional method. Explain why a negative control should or should not be used. Describe, including a species name, an appropriate negative control.

5. Using ELISA method:
   (a) Fill in Table 9.5 with the absorbance readings for the food samples, Lab-positive, Kit-positive, and Kit-negative.
   (b) Calculate the cutoff value; determine and record the presumptive result.
   (c) The ELISA-based method uses a kit that tests for *Salmonella* spp. Could the kit be modified to detect a specific *Salmonella* serovar (e.g., *Salmonella enteritidis*)? Explain.

6. Using the DNA probe method:
   (a) Fill in Table 9.6 with the absorbance readings for the food samples, Lab-positive, Kit-positive, and Kit-negative.
   (b) Determine and record the presumptive result.
   (c) Ribosomal RNA was the target in the DNA probe technique for detection of *Salmonella*. Provide two reasons why this is a logical target.

7. For each of the three methods (conventional, ELISA, and DNA probe), compare and contrast the presumptive group results and class results for the same food.

8. What is the most likely source (or sources) of *Salmonella* in the food tested in these laboratory exercises? How did processing this food fail to eliminate the contamination?

CHAPTER 10

Escherichia coli O157:H7

MICROBIAL TYPING; ALTERNATIVE ELISA METHOD

INTRODUCTION

Properties

Escherichia coli belongs to the family Enterobacteriaceae. It is a gram-negative flagellated rod-shaped bacterium. The microorganism is a facultative anaerobe (i.e., it possesses both respiratory and fermentative metabolic pathways), oxidase negative, and indole positive and does not utilize citrates. The microorganism ferments glucose and other carbohydrates, producing acid and gas. Metabolism of glucose results in pyruvate, which is converted to lactic, acetic, and formic acids. Further metabolism of formic acid results in equal amounts of CO_2 and H_2.

Escherichia coli includes a large number of physiologically diverse strains. Some of this diversity is quite evident when *E. coli* O157:H7 is compared to other strains. Although most strains ferment sorbitol within 48h of incubation at 35–37°C, *E. coli* O157:H7 does not ferment this carbohydrate under these conditions. Most *E. coli* strains grow well at 44.5°C in *E. coli* broth, but O157:H7 does not grow at this temperature. Production of β-glucuronidase is common among *E. coli* strains, but *E. coli* O157:H7 does not produce this enzyme. The substrate utilized by β-glucuronidase is 4-methylumbelliferyl-β-D-glucuronide (MUG). Because *E. coli* O157:H7 does not express the genes which encode β-glucuronidase, this serotype is described as MUG negative. *Escherichia coli* O157:H7 and other members of the enterohemorrhagic group produce cytotoxic factors known as verotoxins, which are now commonly referred to as Shiga-like toxins, or simply Shiga toxins.

Classification

Escherichia coli strains are commonly described serologically by identifying two surface components: the O antigen of lipopolysaccharide (LPS) layer and the flagella antigen (H). There are 167 O antigens and 53 H antigens of *E. coli* that are currently known. To use these antigens for identifying the *E. coli* serotype, two numbers follow the name; the first indicates the O antigen and the second refers to the H antigen. The following are examples of this serotype designation: *E. coli* O55:H6, *E. coli* O111:H8, *E. coli* O157:H7 and *E. coli* O167:H4.

The Disease

Escherichia coli is a natural inhabitant of the intestinal tract and most strains are harmless to humans. Pathogenic *E. coli* strains can be grouped by virotyping. In the virotyping system, strains are grouped based on their virulence factors, clinical syndromes, and mechanisms of pathogenicity. Most pathogenic strains are grouped under these virotypes: enteropathogenic, enterotoxigenic, enteroinvasive, enteroaggregative, and enterohemorrhagic *E. coli* (EHEC). Although there is no clear association between strain virulence and its antigenic makeup, some *E. coli* serotypes are commonly associated with particular virotypes. *Escherichia coli* O157:H7, for example, is a serotype with predominance in the enterohemorrhagic group.

Escherichia coli O157:H7 and other members of the enterohemorrhagic group (e.g., O26:H11, O111:H8, and the nonmotile O157:NM) cause hemorrhagic colitis in humans. The disease mechanism is probably a noninvasive infection. The infective dose is believed to be less than 100 cells. Incubation period is typically 3–4 days during which the microorganism colonizes the large intestine and produces Shiga toxins. Children less than 5 years of age and people older than 65 are the individuals at most risk for infection as well as death from complications. Symptoms of hemorrhagic colitis include severe cramping and diarrhea which begins watery and becomes very bloody as the disease progresses. Vomiting occurs occasionally and fever is rare. In children, hemolytic uremic syndrome (HUS) is a common complication. This leads to renal failure, permanent loss of kidney function, and death. In adults, the primary complication is thrombotic thrombocytopenic purpura (TTP), which resembles HUS histologically but is also accompanied by neurological abnormalities resulting from clots in the brain. Even with the severity of complicating syndromes, mortality among those who contact the disease is less than 5% in North America.

Foods Implicated

Cattle are the main reservoir of *E. coli* O157:H7; therefore, the pathogen is most commonly associated with undercooked ground beef. *Escherichia coli* O157:H7 has been found in a variety of other foods, including raw and pasteurized milk, cheese curds, unpasteurized apple cider and juice, lettuce, alfalfa sprouts, and venison jerky. Drinking water is also a vehicle for transmission of the disease. Outbreaks of the disease associated with *E. coli* O157:H7 peak during the warmest months of the year. Several factors may contribute to the seasonal pattern of disease outbreaks. There may be an increased prevalence of the pathogen in animals during these months. There may be a greater exposure to ground beef (source of *E. coli* O157:H7) as well as the likelihood for greater food abuse in the summer months.

Escherichia coli O157:H7 has been shown to survive in environments where the pH is quite acidic. For example, *E. coli* O157:H7 has been isolated from mayonnaise and apple cider. Both of these foods have pH values of ~3.6. *Escherichia coli* O157:H7 is partially inhibited by 4.5% NaCl in broth medium, although cells have been shown to survive in media where the salt content was 8.5%.

Detection

Conventional methods are frequently used for the detection of *E. coli* O157:H7 in food. These methods rely on cultural, biochemical, and serological tests. Alternative rapid immunology-based methods are available in different formats. These rapid methods, however, require an enrichment step and only negative results are considered conclusive. The following is an overview of procedures used in conventional detection methods.

Enrichment Because the infective dose for *E. coli* O157:H7 is very low, there is a zero-tolerance policy for the presence of *E. coli* O157:H7 in foods. Therefore, presence of one *E. coli* O157:H7 cell in the analyzed food sample deems the food unfit for consumption. Since *E. coli* O157:H7 is usually found in low numbers in foods, an enrichment step is necessary to increase the population of the pathogen to a detectable level. Presence of a large microbiota in meat makes detecting *E. coli* O157:H7 in these products a difficult task. The masking effect of background microbiota can be reduced by using an enrichment with a strong selectivity. Unfortunately, strong selectivity of enrichment broth may also affect the recovery of stressed or injured *E. coli* O157:H7. Therefore, suitability of different enrichment broths for detecting this pathogen in a particular food should be assessed carefully.

Enriching food samples in *E. coli* O157:H7 involves mixing the sample with a rich medium containing selective agents and incubating the mixture. Antibiotics such as cefixime, cefsulodin, novobiocin, and vancomycin are used to select for the EHEC in this medium.

Isolation Although there is no obvious relationship between *E. coli* serotypes and biotypes, some serotypes can be separated based on their metabolic differences. Isolation of the *E. coli* O157:H7 from the enrichment includes plating on agar media that select for *E. coli* and differentiate colonies based on sorbitol fermentation and MUG reaction. Sorbitol–MacConkey (SMAC) agar is commonly used for isolating *E. coli* O157:H7 based on inability of the bacterium to ferment sorbitol. When potassium tellurite and cefixime are added to SMAC, the resulting medium (TC SMAC) improves the recovery of the pathogen from food. Analysts should be cautioned, however, that there are non–*E. coli* O157:H7 isolates that do not ferment sorbitol and produce colonies on SMAC and TC SMAC that look similar to those of *E. coli* O157:H7. Therefore, sorbitol-negative isolates should be subsequently screened by the testing for the MUG reaction. Alternatively, hemorrhagic colitis (HC) agar is used for isolation. This medium differentiates isolates based on both sorbitol fermentation and MUG reaction.

Confirmation

- Sorbitol-negative, MUG-negative isolates must be confirmed as *E. coli* using the indole test and other suitable biochemical assays. If the isolate is indole negative, it is not considered to be an *E. coli* and thus should be discarded.
- Sorbitol-negative, MUG-negative *E. coli* is then tested serologically for the presence of O157 or both O157 and H7 antigens. This can be accomplished using commercial O157 or O157 and H7 antisera. Several commercial test kits are available for this confirmation (see Meng et al., 2001). Isolates that produce positive antigen–antibody interaction are considered *E. coli* O157 or *E. coli* O157:H7. Instead of serotyping using antisera, a DNA hybridization or a PCR technique with probes that target gene encoding for O157 and H7 antigens may be used.
- Additional confirmatory tests may be needed to prove that the isolate is capable of producing the Shiga toxin. Such tests include toxicity to HeLa and Vero tissue culture cells or genetic techniques that detect genes encoding for the Shiga toxins.

OBJECTIVES

1. Detect *E. coli* O157:H7 in food.
2. Identify the pathogenic strain based on its biochemical and serological characteristics.
3. Practice a rapid test using an immunochromatographic technique.

PROCEDURE OVERVIEW

Escherichia coli O157:H7 will be detected in food by a method adapted from Cray et al. (1998) and developed in compliance with the principles presented in the introduction of this chapter (Fig. 10.1). Food will be enriched in *E. coli* O157:H7 using the modified EC broth with novobiocin. Samples are mixed with the enrichment broth and incubated at 35°C for 24 h. Enrichment is transferred with a swab onto Rainbow agar plates and plates are incubated at 35°C for 24 h. Colonies of *E. coli* that lack β-glucuronidase (MUG negative) appear black or grey on the Rainbow agar; these are considered presumptive *E. coli* O157:H7. Presumptive colonies are then tested for the presence of the O157 antigen using a commercial test kit (NOW, Binax, Portland, ME). Additionally, these colonies are confirmed as sorbitol negative by incubation in phenol red sorbitol broth. The presumptive colonies are also tested for indole production; *E. coli* is indole positive. For a complete identification of the isolate as EHEC, additional biochemical confirmation is needed, and production of Shiga toxin by the isolate should be verified. These additional tests, however, will not be completed in this laboratory exercise.

Results of this exercise will be interpreted as follows. Colonies that are black or grey on Rainbow agar (MUG negative), positive for O157 antigen as analyzed by the commercial test kit, sorbitol negative in phenol red sorbitol broth, and indole positive will be considered as *E. coli* O157:NM (nonmotile) or *E. coli* O157:H7.

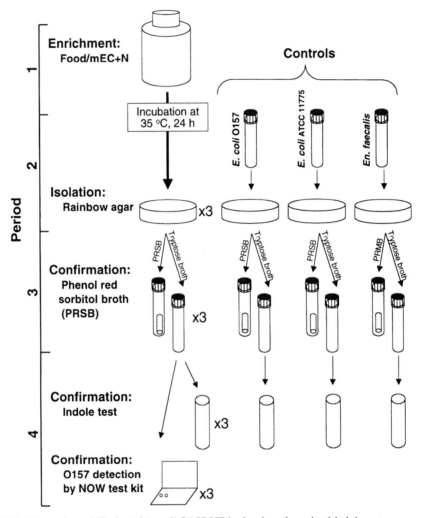

Fig. 10.1. Detection of *Escherichia coli* O157:H7 in food as done in this laboratory exercise.

Detection of O157 Antigen by Immunochromatographic Assay

The commercial kit (NOW) will be used to detect the O157 antigen of *E. coli* in broth containing presumptive *E. coli* O157:H7 isolates (Fig. 10.2). This is an immunochromatographic test which helps the analyst visualize conjugated antibodies that bind target antigens as they flow across a membrane. The basis for the assay is very similar to that of the enzyme-linked immunosorbent assay (ELISA), which was presented in Chapter 9.

The test kit swab is soaked in the broth that may contain *E. coli* O157:H7. The swab is then inserted in a compartment in the test card. An extracting reagent is then added to the swab tip in the upper hole of the card. As the sample flows up the test strip, the O157 antigens present will bind to the rabbit anti-O157 antibody (conjugated to a visualizing particle) precoated on the test membrane. As the antigen–antibody complex continues to flow up the test strip, it will bind to anti-

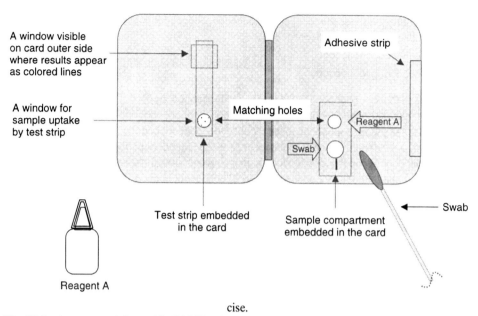

Fig. 10.2. A commercial test kit (NOW, Binax, Portland, ME) for immunochromatographic detection of O157 antigen in samples enriched in *E. coli* O157:H7.

O157 antibodies that are adhering to a test line. When these antibodies are bound, a color is produced and the line can be visualized. An internal test control of conjugated goat immunoglobulin G (IgG) is also present in the test. This conjugate will bind to the immobilized rabbit anti-goat antibody present on the control line located at the top of the visualizing window. This control aids in determining the validity of the test. Kit-positive and kit-negative controls are also included.

Indole Test

Escherichia coli, including the EHEC strains, decomposes tryptophan in the growth medium and produces indole. Peptone is one of the sources of tryptophan in microbiological media. Production of indole can be detected using Kovac's indole reagent (*p*-dimethylamine-benzaldhyde, dissolved in butanol–HCl). When *E. coli* liquid culture is mixed with Kovac's reagent, the presence of a red color in the upper (alcohol) layer indicates an indole-positive culture.

MEDIA

Modified *E. coli* Broth with Novobiocin (mEC–N)

The EC broth is a buffered lactose medium that may be used to differentiate and enumerate coliforms. In this laboratory, it is used to reduce the number of contaminating noncoliforms. This medium contains lactose, which is fermented by coliforms to acid and gas. It also contains bile salts, which inhibit the growth of gram-positive bacteria. This medium is modified by adding novobiocin, which further suppresses the growth of gram positives and some gram negatives. Most *Enterobacteriaceae* organisms are resistant to novobiocin.

Phenol Red Sorbitol Broth

Phenol red carbohydrate broth is a basal medium used to differentiate bacterial isolates based on the fermentation of the added carbohydrate. The medium contains peptone and beef extract as sources of carbon and nitrogen, phenol red as a pH indicator, and a selected carbohydrate. When the carbohydrate is fermented creating acid end products, the pH indicator in the medium changes from red to yellow. When gas is also produced, it is entrapped and detected in Durham tubes. When sorbitol is added as a fermentable carbohydrate, the medium becomes suitable for differentiating and thus confirming presumptive *E. coli* O157:H7 since this serotype is typically sorbitol negative. Other *E. coli* strains ferment sorbitol to acidic end products.

Rainbow Agar

Rainbow agar (Biolog, Hayward, CA) is a selective and differential chromogenic medium that inhibits the growth of noncoliforms while distinguishing between biotypes of *E. coli*. According to the manufacturer, tellurite can be added to reduce background microbiota and select for *E. coli* O157:H7. Novobiocin is added to inhibit *Proteus* swarming and the growth of tellurite-reducing bacteria. Rainbow agar also contains chromogenic substrates that detect the production of β-galactosidase and β-glucuronidase. Presence of blue-black colonies on Rainbow agar indicates β-galactosidase activity. β-Glucuronidase activity is indicated by a red color. *Escherichia coli* O157:H7 lacks β-glucuronidase and forms unique black or grey colonies. Other *E. coli* produce colors that depend on the relative production of β-galactosidase to β-glucuronidase. Nearly all non–*E. coli* species are either inhibited or produce white or cream-colored colonies.

Tryptose Broth

This is a rich nonselective medium that will be used in this exercise to transfer presumptive *E. coli* O157:H7 isolates and non–*E. coli* control culture, and as a vehicle for the indole test.

ORGANIZATION

Students will work in groups of two. Each pair of students will test one food sample. Ground beef, from different sources, including local small butcher shops, may be tested.

PERSONAL SAFETY

Escherichia coli O157:H7 is a highly pathogenic microorganism that should be handled with care. Follow the safety guidelines that were reviewed earlier in this manual. Use disposable gloves when handling plates and cultures. Make sure the work area is sanitized carefully after each use.

REFERENCES

Binax/NEL NOW EH E. coli *O157 and O157:H7 Brochure*. 1999. Binax, Portland, ME.

Cray, W. C., D. O. Abbott, F. J. Beacorn, and S. T. Benson. 1998. Detection, Isolation and Identification of *Escherichia coli* O157:H7 and O157:NM (Nonmotile) from Meat Products. In *USDA-FSIS Microbiology Laboratory Guidebook*, 3rd ed., rev. 2 (February 23, 2001). U.S. Department of Agriculture, Washington, DC.

Hitchins, A. D., P. Feng, W. D. Watkins, S. R. Rippey, and L. A. Chandler. 2001. *Escherichia coli* and the Coliform Bacteria. In *US FDA-CFSAN, Bacteriological Analytical Manual Online* (Chapter 4). Available: http://cfsan.fda.gov/~ebam/bam_toc.html.

Meng, J., M. P. Doyle, T. Zhao, and S. Zhao. 2001a. Enterohemorrhagic *Escherichia coli*. In M. P. Doyle, L. R. Beuchat, and T. J. Montville (Eds.), *Food Microbiology, Fundamentals and Frontiers*, 2nd ed. (pp. 193–231). American Public Health Association, Washington, DC.

Meng, J., P. Feng, and M. P. Doyle. 2001b. Pathogenic *Escherichia coli*. In F. P. Downes and K. Ito (Eds.), *Compendium of Methods for the Microbiological Examination of Foods*, 4th ed. (pp. 331–341). American Public Health Association, Washington, DC.

Rainbow Agar O157 Technical Information. 1999. Biolog, Hayward, CA.

Period 1 Enrichment

MATERIALS AND EQUIPMENT

Per Pair of Students

- Food sample, 25 g
- Modified EC broth with novobiocin (mEC–N), 225 ml

Class Shared

- Scale for weighing food samples (e.g., a top-loading balance with 500 g capacity)
- Incubator, set at 35°C
- Stomacher and stomacher bags
- Tub or large glass beaker for holding enrichment bags

PROCEDURE

1. Prepare the ground meat to ensure that a representative sample is analyzed. This includes mixing the package contents in a sterile container before taking a sample.
2. Weigh 25 g of the ground meat into a stomacher bag. Add 225 ml mEC–N to the bag. Stomach sample for 2 min.
3. Place the bag in the tub or beaker designated by the laboratory instructor. This container should support the bag and keep it upright during incubation and handling.
4. Incubate the enrichments at 35°C for 24 h.

Period 2 Isolation

MATERIALS AND EQUIPMENT

Per Pair of Students

- Incubated enrichment broth
- Six Rainbow agar plates (with novobiocin)
- *Escherichia coli* O157 (lab-positive control) in EC broth
- *Escherichia coli* ATCC 11775 or equivalent (a nonpathogenic lab-negative control) in EC broth
- *Enterococcus faecalis* (a non–*E. coli* lab-negative control) in tryptose broth
- Disposable gloves
- Three sterile swabs

Class Shared

- Incubator, set at 35°C

PROCEDURE

Control

1. Label three Rainbow agar plates for the three controls.
2. Use the inoculation loop to prepare three-phase streaks from the broth of the control cultures.
3. Incubate all plates at 35°C for 24 h.

Food Sample

1. Label three Rainbow agar plates appropriately for the food sample.
2. Carefully mix the contents of the stomacher bag.
3. Moisten a swab in the enrichment culture, being careful not to take large food particles. Use the swab to make a primary streak on a Rainbow agar plate.
4. Using an inoculation loop, complete two additional streak phases.
5. Repeat steps 3 and 4 using separate swabs to inoculate two additional Rainbow agar plates from the enrichment in the stomacher bag.
6. Incubate all plates at 35°C for 24 h.

Period 3 Isolation (Continued) and Confirmation

MATERIALS AND EQUIPMENT

Per Pair of Students

- Incubated Rainbow agar plates (from the previous laboratory period)
- Six phenol red sorbitol broth tubes with Durham tubes, 9 ml each
- Six tryptose broth tubes, 9 ml each
- Disposable gloves

Class Shared

- Incubator, set at 35°C
- Colony counter

PROCEDURE

Control

1. Examine the incubated Rainbow agar control plates. Note the color of each culture in Table 10.1.
2. Choose a well-isolated colony from each control plate. Transfer a portion of the colony into a phenol red sorbitol tube using inoculation loop. Be careful not to disturb the inverted Durham tube at the bottom of the test tube (i.e., mix gently and do not vortex the tube contents).
3. Transfer a part of the same colonies from control plates into three tryptose broth tubes using an inoculation loop. Mix the contents of the tube.
4. Incubate all tubes at 35°C for 24 h.

Food Sample

1. Examine the incubated Rainbow agar sample plates. Identify any black or grey colonies, which indicate presumptive *E. coli* O157:H7 in food. Record the observations in Table 10.1.

TABLE 10.1. *(Add a descriptive title, including food and how results were obtained)*

Sample	Observations
Food	
E. coli O157:H7	
E. coli ATCC 11775	
E. faecalis	

(Add footnotes as required for this table).

2. Choose three presumptive *E. coli* O157:H7 colonies for confirmation.
3. Transfer portions of the three presumptive colonies into three phenol red sorbitol tubes using an inoculation loop. Mix gently (do not vortex) the tube contents to avoid disturbing the Durham tube.
4. Transfer the remaining portions of the same three presumptive colonies into three tryptose broth tubes using an inoculation loop. Mix the contents of the tube.
5. Incubate all tubes at 35°C for 24 h.

Period 4 Confirmation (*Continued*)

MATERIALS AND EQUIPMENT

Per Pair of Students

- Incubated phenol red sorbitol broth tubes receiving controls (three tubes)
- Incubated tryptose broth tubes receiving controls (three tubes)
- Incubated phenol red sorbitol broth tubes receiving food sample isolates (three tubes)
- Incubated tryptose broth tubes receiving food sample isolates (three tubes)
- Six test cards (NOW commercial test kit)
- Six test swabs (NOW commercial test kit)
- Disposable gloves

Class Shared

- Kovac's indole reagent
- Reagent A (NOW commercial test kit)

PROCEDURE

Sorbitol Fermentation

1. Examine the incubated phenol red sorbitol broth tubes. Check tubes for sorbitol fermentation. If sorbitol is fermented, the tubes become yellow.
2. Check for gas entrapment in the Durham tubes.
3. Record the observations in Table 10.2.

TABLE 10.2. (*Add a descriptive title, including food and how results were obtained*)

Isolate Source	Sorbitol Fermentation	Indole Production	Presumptive *E. coli* O157:H7?
Food			
Food			
Food			
E. coli O157:H7			
E. coli ATCC 11775			
E. faecalis			

(*Add footnotes as required for this table*).

Indole Production

1. Carefully transfer 5 ml of each incubated tryptose broth tube into an empty clean test tube.
2. Add 0.5 ml Kovac's indole reagent to each of the newly prepared test tubes. Do not mix.
3. Observe whether a red color appears in the upper (reagent) layer. A red layer indicates a positive indole reaction, while a yellow/green color is a negative reaction.
4. Record your observation in Table 10.2.

Detection of O157 Antigen

A commercial test kit (NOW) will be used in this analysis, as previously described (Fig. 10.2). Use the kit's positive and negative controls (if available) instead of the laboratory-prepared controls. Read carefully the kit inserts for a complete description of the test and check for any procedural changes introduced by the kit producer.

Isolates from Food Samples

1. Open a test card (commercial test kit) and label the front cover with the type of sample analyzed. Notice the two holes on the inner right panel.
2. Obtain a foam swab (commercial test kit).
3. Immerse the foam swab into the remainder of the trypose broth culture. If there is too much liquid on the swab, just roll the swab against the inside of the test tube once or twice.
4. Insert the moistened foam swab into the bottom hole of the test card and gently push it upward until the tip rests in the top hole.
5. Add 2 drops of reagent A to the top hole (hold the vial vertically ~1 inch above the top hole).
6. Remove the paper strip from the right-hand side of the card to expose the adhesive strip and close the card.
7. Read the test results as lines that appear in the clear window on the outer side of the card. Results should be read within 10 min.
8. Repeat the previous steps for the remaining food samples.
9. Record results in Table 10.3.
10. If no kit controls are available, repeat steps 1–9 with the laboratory-prepared controls.

Kit Controls

1. Open a test card (commercial test kit) and label the front cover with the type of control analyzed.
2. Obtain the foam swab representing the negative control (commercial test kit).
3. Insert the foam swab into the bottom hole of the test card and gently push it upward until the tip rests in the top hole.

TABLE 10.3. *(Add an appropriate title, including food and testing methods)*

Isolate Source	NOW Test Kit Result
Food	
Food	
Food	
Kit's positive control	
Kit's negative control	

(Add footnotes as required for this table).

TABLE 10.4. Interpretation of Results Obtained in Laboratory Exercise to Detect *Escherichia coli* O157:H7 in Food

Test and Medium	Result	Interpretation
Isolation on Rainbow agar	Black or grey colonies	β-Glucuronidase not produced
Sorbitol fermentation (phenol red sorbitol broth)	Negative (no color change, no gas produced)	Sorbitol not fermented
Detection of O157 antigen (NOW, a commercial test kit)	Positive	O157 antigen present
Indole test (Kovac's reagent)	Positive	Isolate confirmed as *E. coli*

4. Add 2 drops of reagent A to the top hole (hold the vial vertically ~1 inch above the top hole).
5. Remove the paper strip from the right-hand side of the card to expose the adhesive strip and close the card.
6. Read the results within 10 min.
7. Repeat the previous steps for the kit's positive control swab.
8. Record the results in Table 10.3.
9. Interpret the results as follows:
    a. The negative control swab will result in a line at the top half of the test window only.
    b. The positive control swab will result in a line in the top half and another in the bottom half of the window.
    c. Results are invalid if no line appears in the control area or if only the sample line is visible.
    d. Sample swabs are interpreted by matching these results to the negative or positive control swab.

Data Interpretation

Results obtained in this laboratory exercise can be interpreted as indicated in Table 10.4. Although these results can help the analyst conclude whether *E. coli* O157:H7 is present or absent in food, these data are not sufficient to determine if the isolates cause hemorrhagic colitis in humans. Additional tests to confirm the isolate's identity and ability to produce Shiga toxin are needed before such conclusions can be made. It should be kept in mind, however, that some cases of hemorrhagic colitis are caused by *E. coli* serotypes other than O157:H7.

PROBLEMS

1. *Escherichia coli* O157:H7 can be found in many foods. What foods should be analyzed for this pathogen more frequently than others? Why?

2. For the food analyzed in this laboratory exercise, list the following information:
   (a) Method of packaging
   (b) Storage method
   (c) Presence of preservative, if any

   Describe how each of the above factors influences the microbial load of the food.

3. Complete Tables 10.1–10.3 with the data gathered in this laboratory exercise. For each table, include a suitable title and a footnote.

4. Explain why three control cultures were used in this laboratory. Include what each contributed and if they were equally useful in all tests.

5. The commercial test kit used in this exercise (NOW) detects O157 antigen. Explain the benefit and limitations of using only this antigen in the test.

6. List a recent outbreak of hemorrhagic colitis caused by *E. coli* O157:H7. How would the analysis you learned in this exercise be useful in tracking the cause of such an outbreak?

7. Suppose you decided to develop a DNA probe that targets all enterohemorrhagic strains of *E. coli*. What gene or genes would you use to develop such a probe? Explain.

PART IV

FOOD FERMENTATION

INTRODUCTION

The presence of microorganisms in many foods is undesirable since most microorganisms spoil the food or render it unsafe for human consumption. Food preservation technologies target these microorganisms by elimination, inactivation, or inhibition. Elimination includes exclusion by aseptic processing and proper packaging or removal by microfiltration. Microbial inactivation (killing) is accomplished by physical treatments such as heat or application of chemical preservatives (e.g., acids) or sanitizers (e.g., chlorine). Inhibiting growth of microorganisms in food is another commonly used preservation strategy. Most chemical preservatives (e.g., sorbates and nitrites) retard microbial growth and thus extend the shelf life of food.

Some microorganisms are considered beneficial. These are used in certain foods and in non-food-related industrial fermentations. Fermentation is broadly defined as the biochemical changes in organic substances that are caused by microbial enzymes. The presence of metabolically active microorganisms is therefore essential for fermentation processes. Some of the familiar fermentations include using yeasts in bread-making, production of alcoholic beverages, and conversion of agricultural commodities such as corn into fuel ethanol. Other fungi are used in production of mold-ripened cheese or production of important pharmaceuticals, such as penicillin. Lactic acid bacteria (LAB) carry out fermentations leading to the production of yogurt, cheese, sausage, and pickles. Use of LAB in food fermentation will be addressed in this part of the manual.

Fermentations lead to the production of new and popular food varieties. Milk, for example, can be made into more than 400 cheese varieties, many of which result from varying the fermentation process. Most food fermentations also lead to preservation. Fermentation of food by LAB, for example, results in the production of lactic acid, which lowers the pH of the product and makes it unsuitable for growth of spoilage or pathogenic microorganisms. The large population of LAB in fermented foods competes strongly for available nutrients with microbial contaminants and thus enhances product safety. Some LAB produce antimicrobial peptides, known as bacteriocins, which may target certain pathogens. Nisin is an example of a bacteriocin produced by LAB; this bacteriocin is active against several spoilage and pathogenic microorganisms.

LACTIC ACID BACTERIA

Properties

Lactic acid bacteria constitute a heterogeneous group of microorganisms with some phenotypic similarities. These are gram-positive, non-spore-forming bacteria that lack cytochromes and other heme-based compounds (e.g., true catalase). Consequently, LAB are microaerophilic or anaerobic but are air tolerant and metabolize carbohydrates through fermentative pathways. These bacteria are fastidious; hence they are associated with foods rich in nutrients such as milk and meat. Most LAB are acid tolerant and some are also salt tolerant.

Genera under LAB differ in cell morphology; some are cocci and others are rods. Although most LAB grow well in the mesophilic temperature range, some show psychrotrophic properties (grow at 10°C but not at 45°C), and others tend to be thermophiles (grow at 45°C but not at 10°C). Genera of LAB vary in the mode of glucose fermentation; some LAB ferment the sugar, producing lactic acid only (homolactics), whereas others produce alcohol and carbon dioxide in addition (heterolactics). Carbohydrate fermentation by LAB yields D-, L-, or both isomers of lactic acid. These differences help in differentiating the genera under LAB group (Table IV.1).

Genera of Interest

Ten genera, at least, of gram-positive bacteria are considered LAB: *Corynebacterium*, *Enterococcus*, *Lactococcus*, *Lactobacillus*, *Leuconostoc*,

TABLE IV.1. Properties of Selected Genera of LAB Group[a]

Property	*Lactobacillus*	*Lactococcus*	*Streptococcus*	*Pediococcus*	*Leuconostoc*
Morphology	Rods	Cocci	Cocci	Cocci (forms tetrads)	Cocci
CO_2 from glucose	±	−	−	−	+
Growth					
At 10°C	±	+	−	±	+
At 45°C	±	−	±	±	−
Relevant species	*bulgaricus, acidophilus*	*lactis, cremoris*	*thermophilus*	*acidilactici*	*mesenteroides*
Miscellaneous	Diverse genus	Lancefield group N[b]	Some are pathogenic	Some grow at 6.5% salt	Some grow at 6.5% salt

[a] ±: species vary.
[b] Lactococci have group N antigen, as defined in Lancefield serotyping scheme.

Oenococcus, Pediococcus, Streptococcus, Tetragenococcus, and *Vagococcus.* The following are genera of particular interest in food fermentation.

Lactobacillus Lactobacilli are rods, usually in long chains. Lactobacilli includes homolactic and heterolactic species. Some lactobacilli cause food spoilage (e.g., *Lb. plantarum*), whereas others are valuable starter cultures in the dairy industry (e.g., *Lb. bulgaricus* and *Lb. helveticus*).

Lactococcus Cells are cocci, found singly, in pairs or as chains. *Lactococcus* spp. are homolactic, grow at 10°C but not at 45°C, and react with group N antisera (Lancefield group N). Members of this genus are frequently used as starter cultures for food applications. Dairy products manufacturers often use *Lc. lactis* subsp. *lactis* and *Lc. lactis* subsp. *cremoris* in cheese making.

Streptococcus Cells are similar morphologically to *Lactococcus* spp. and are also homolactics. Most species grow well at 45°C but not at 10°C. *Streptococcus thermophilus* is used as starter culture for making yogurt and cheeses, particularly the Italian varieties.

Pediococcus These are cocci that exist singly, in pairs or as tetrads. Pediococci are homolactic bacteria that play an important role in meat fermentation. Useful pediococci include *P. acidilactici* and *P. cerevisiae*.

Leuconostoc Leuconostocs are cocci that can be found in pairs or chains. These are heterolactic bacteria that ferment carbohydrates, producing acid and gas. Leuconostocs are relatively acid resistant bacteria with moderate salt tolerance. Species under this genus are useful in vegetable fermentations (e.g., *Leu. mesenteroides*).

Acid Production

Lactic acid production is associated with LAB primary metabolism, and thus actively growing culture produces more acid than do resting cells. When glucose is not limited in the growth medium, growth factors are present, and oxygen level is limited (i.e., microaerophilic conditions), LAB metabolize this sugar by one of the following pathways:

(a) Homolactic fermentation (Embden-Meyerhof-Parnas pathway):

$C_6H_{12}O_6 \rightarrow 2\ CH_3\text{-}CH(OH)\text{-}COOH + 2\ ATP$ (adenosine triphosphate)
(Glucose is converted into lactic acid, almost quantitatively.)

(b) Heterolactic fermentation (phosphoketolase pathway):

$C_6H_{12}O_6 \rightarrow CH_3\text{-}CH(OH)\text{-}COOH + CH_3\text{-}CH_2OH + CO_2 + ATP$
(Glucose is converted into lactic acid, ethanol, and carbon dioxide gas.)

If oxygen concentration is not limited (i.e., conditions are aerobic), acetic acid, instead of ethanol, is produced by heterolactic LAB:

$C_6H_{12}O_6 \rightarrow CH_3\text{-}CH(OH)\text{-}COOH + CH_3\text{-}COOH + CO_2 + 2\ ATP$

Bacteriocin Production

Bacteriocins are peptides produced by some bacteria that inhibit closely related species. These antimicrobial peptides may exhibit bactericidal or bacteriostatic action against targeted microorganisms. The mode of action (i.e., bacteriostatic vs. bactericidal) depends on the complexity of medium (buffer vs. nutritious medium), treatment conditions (e.g., refrigeration vs. optimum growth temperature), physiological status of the microorganism (actively metabolizing vs. resting cells), and presence of media ingredients with affinity for the bacteriocin. When compared with antibiotics, bacteriocins are generally narrower in activity spectrum. Bacteriocins are believed to be membrane-active agents that create channels through cytoplasmic membranes and dissipate proton-motive force, leading to cell death. Several bacteriocins are currently known (e.g., colicin of *E. coli*), but those produced by LAB are of particular importance to the food industry. It has been shown that bacteriocins produced by some LAB are active against foodborne pathogens. Addition of these bacteriocins or the producer strain to food is potentially useful in controlling these pathogens. The safety of fermented foods may be partially attributed to the presence of such bacteriocins.

A few LAB strains only produce bacteriocins. Table IV.2 includes a partial list of LAB bacteriocins of potential importance to the food industry. Nisin, pediocin, lacticin, and thermophilin are some of the bacteriocins produced by LAB. Currently, nisin is the most widely used bacteriocin in food applications.

Nisin

Nisin is a polypeptide with 34 amino acid residues. The bacteriocin has antimicrobial activity against *Clostridium botulinum*, *Listeria monocytogenes*, and other gram-positive bacteria. It prevents the outgrowth of germinating spores. Nisin is nontoxic when consumed with food, and it is degraded by proteolytic enzymes during digestion. Nisin is relatively heat stable; therefore, it may be used in thermally processed foods as a heat adjunct; this application allows processors to use less heat in food processing. Nisin is more effective as an antimicrobial agent in acid than in low-acid foods. Numerous countries, including the United States, allow using nisin as a food additive.

Biopreservation

Biopreservation refers to the use of beneficial bacteria or fermentation products produced by these bacteria in controlling spoilage and pathogenic microorganisms in food. The preserving action of these bacteria is not necessarily associated with their ability to ferment the food of interest. Bacteria used in biopreservation should be harmless to humans (e.g., members of a LAB group), compete well with spoilage and pathogenic microorganisms for nutrients in food if conditions are conducive to microbial growth, and produce acids and other antimicrobial agents, particularly bacteriocins.

MICROBIAL GROWTH AND GROWTH KINETICS

Microorganisms increase in number or mass when present in an environment that supports their growth and when permitted in this environment for a suitable length

TABLE IV.2. Selected Bacteriocins of LAB and Their Inhibitory Spectra

Bacteriocin	Producer	Inhibitory Spectrum		Reference
		Pathogens	Nonpathogens	
Nisin	*Lactococcus lactis* subsp. *lactis* ATCC 11454, JS21, NIZO R5, INRA 1-6, NP4G, NZI, NCDO2111, NCDOS1, ILC13	*Bacillus cereus, Clostridium perfringens, Listeria monocytogenes, Staphylococcus aureus, Aeromonas hydrophila, Escherichia coli* O157:H7, *Vibrio cholerae, Vibrio parahaemolyticus, Enterococcus faecalis*	*Listeria innocua, L. ivanovii, L. seeligeri, L. welshimeri, L. murrayi, Micrococcus flavus, M. luteus; Streptococcus thermophilus; Clostridium butyricum, C. sporogenes, C. tyrobutyricum; Lactococcus lactis* subsp. *cremoris, Lc. diacetylactis; Lactobacillus leichmannii, Lb. plantarum; Pediococcus acidilactici*	DeVos et al., 1993; Leung et al., 2002; Meghrous et al., 1999; Spelhaug and Harlander, 1989
Pediocin AcH	*Pediococcus acidilactici* H	*Aeromonas hydrophila, Bacillus cereus, Clostridium perfringens, Listeria monocytogenes, Staphylococcus aureus*	*Lactobacillus plantarum, Lb. viridecens; Leuconostoc mesenteroides; Pseudomonas putida*	Bhunia et al., 1988
Plantaricin S	*Lactobacillus plantarum* LPCO10	*Enterococcus faecalis*	*Clostridium tyrobutyricum; Propionibacterim* sp.; *Lactococcus lactis* subsp. *cremoris, Lc. lactis* subsp. *lactis; Leuconostoc mesenteroides, Leu. paramesenteroides; Lactobacillus delbrueckii, Lb. fermentum, Lb. helveticus, Lb. sake, Lb. curvatus; Micrococcus* sp.*; Pediococcus pentosaceus; Streptococcus thermophilus*	Jimenez-Diaz et al., 1993
Propionicin PLG-1	*Propionibacterium thoenii*	*Listeria monocytogenes, Vibrio parahaemolyticus, Yersinia enterocolitica*	*Pseudomonas fluorescence; Corynebacterium* sp.	Lyon et al., 1993
Thermophilin 13	*Streptococcus thermophilus*	*Listeria monocytogenes* 59, *Bacillus cereus* C14, *Clostridium botulinum* 100003	*Streptococcus thermophilus; Enterococcus faecium; Lactococcus cremoris; Lactobacillus acidophilus, Lb. helveticus, Lb. fermentum; Leuconostoc cremoris, Leu. mesenteroides; Bifidobacterium bifidum; Listeria innocua; Bacillus subtilis; Clostridium tyrobutyricum; Staphylococcus carnosus; Micrococcus varians*	Marciset et al., 1997

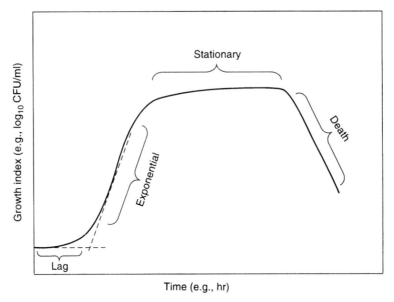

Fig. IV.1. Phases of microbial growth.

of time. A supportive environment includes available water and nutrients, proper level of oxygen (i.e., high, low, or absent), suitable incubation temperature, and absence of inhibitory or microbicidal agents or conditions. When these conditions are met, a small microbial inoculum grows in a characteristic pattern or phase.

Growth Phases

Phases of growth are readily defined when illustrated as a growth curve (Fig. IV.1). The microbial growth curve is a graphical representation of the changes in number, mass, or other suitable growth indices while incubating the microorganism in a suitable medium. The following are the important phases of a typical microbial growth:

1. *Lag Phase* Characteristically, the growth curve includes a "lag phase" or "lag period" during which no signs of growth are observed. In fact, it is not uncommon to observe a slight decline in the population size during this phase. During the lag period, the microorganism adjusts to the new environment and activates suitable metabolic machinery for the best use of nutrients and conditions in this environment.

Optimization of batch fermentation processes may include provisions to decrease the microbial lag phase. This saves time and resources during fermentations. The lag phase is also important to food processors for a different reason. Several preservation technologies are designed to extend the lag period for microorganisms that were introduced into food during harvesting or processing. This strategy aims at increasing product shelf life.

2. *Exponential Phase* After the lag, microorganisms gradually increase their number or mass in an exponential fashion. Since ample supply of nutrients is available and bacteria multiply by binary fission, each bacterial cell divides at equal time

intervals, leading to exponential (2^n) increase in cell number. This time interval is known as generation time or doubling time. In case of mold growth, this parameter may be defined as the time required to double the mass of mycelium. Specific growth rate, a parameter related to generation time, measures rate of growth independent of cell count or mass (see Appendix B for more details). Generation time reaches its smallest value, and specific growth rate is at a maximum value during the exponential phase of growth. When microorganisms are grown under almost ideal conditions, different generation times or specific growth rates are observed. Therefore, these parameters are associated with inherent physiological properties of the microorganism. Environmental factors during fermentation (e.g., temperature and rate of aeration) also affect the generation time.

Generation time (or specific growth rate) is a parameter of great significance in food fermentation and preservation. Food fermentations may be accelerated by adjusting processing conditions to shorten the doubling time. In case of food preservation, however, spoilage and pathogenic microorganisms with short generation time (e.g., *Pseudomonas* spp. and *Clostridium perfringens*, respectively) are troublesome. Preservation technologies that do not eliminate or inactivate the microorganism should at least slow microbial growth in food, that is, extend the generation time of microbial contaminants.

3. *Stationary Phase* Depletion of essential nutrients and accumulation of growth-suppressing microbial metabolites (e.g., acids or alcohols) arrest growth of the microorganism and initiate the stationary phase and subsequently the death phase. The microbial population size reaches its maximum at this phase. Maximum growth varies with many factors, including the inoculated microorganism, medium composition, and environmental factors during culture incubation. When many foodborne bacteria are inoculated in rich medium or food and incubated under ideal conditions for growth, a maximum growth of ~10^9 CFU/ml may be achieved.

4. *Death Phase* Cell death follows the stationary phase. Interestingly, cells die logarithmically and thus first-order kinetics may be applied to this phase.

Measuring Growth Parameters

Lag period, generation time (or specific growth rate), and maximum growth are important parameters that define the lag, exponential, and stationary phases of microbial growth, respectively. To measure these parameters, data about changes in microbial population during incubation should be gathered and preferably presented graphically. This, for example, may involve plotting $\log_{10}$ CFU/ml (Y axis) against time in minutes or hours (X axis). These data may be fitted using nonlinear mathematical functions and growth parameters are computed. Alternatively, simpler and approximate techniques may be used to estimate these parameters. Lag phase and maximum growth values may be determined directly from the plotted growth curve.

Unlike the lag period or maximum growth, the generation time or specific growth rate is not readily measurable from the plotted data. The following are mathematical formulas for these parameters that make use of the plotted growth curves:

Specific growth rate (μ):

$$\mu = \frac{2.3(\log_{10} x_2 - \log_{10} x_1)}{t_2 - t_1} \tag{1}$$

Generation time (GT):

$$GT = \frac{t_2 - t_1}{3.3(\log_{10} x_2 - \log_{10} x_1)} \qquad (2)$$

where x_1 and x_2 are cell populations (e.g., CFU/ml) and t_1 and t_2 are the corresponding incubation times (e.g., minutes) during the exponential phase of growth. Both specific growth rate and generation time describe the exponential phase of growth, although these values are inversely proportional to each other. Therefore, the shorter the generation time, the greater the specific growth rate of the microorganism under a given set of growth conditions. To use these equations in calculating μ or GT, choose a segment of the growth curve that represents the exponential phase (preferably the steepest portion of the curve), determine the coordinates of the two data points that flank this segment (i.e., x_1, t_1 and x_2, t_2), and apply these numbers in Eq. (1) or (2). Derivation of Eqs. (1) and (2) is detailed in Appendix B.

REFERENCES

Axelsson, L. T. 1993. Lactic Acid Bacteria: Classification and Physiology. In S. Salminen and A. von Wright (Eds.), *Lactic Acid Bacteria* (p. 164). Marcel Dekker, New York.

Bhunia, A. K., M. C. Johnson, and B. Ray. 1988. Purification, Characterization and Antimicrobial Spectrum of a Bacteriocin Produced by *Pediococcus acidilactici*. *J. Appl. Bacteriol.* 65:261–268.

DeVos, W. M., J. W. M. Mulders, R. J. Siezen, J. Hugenholtz, and O. P. Kuipers. 1993. Properties of Nisin Z and Distribution of Its Gene, *nis*Z, in *Lactococcus lactis*. *Appl. Environ. Microbiol.* 59:213–218.

Jay, J. M. 2000. *Modern Food Microbiology*. Aspen Publishers, Gaithersburg, MD.

Jimenez-Diaz, R., R. M. Rios-Sanchez, M. Desmazeaud, J. L. Ruiz-Barba, and J. C. Piard. 1993. Plantaricin S and T, Two New Bacteriocins Produced by *Lactobacillus plantarum* LPCO10 Isolated from a Green Olive Fermentation. *Appl. Environ. Microbiol.* 59:1416–1424.

Klaenhammer, T. R. 1993. Genetics of Bacteriocins Produced by Lactic Acid Bacteria. *FEMS Microbiol. Rev.* 12:39–86.

Leung, P. P., M. A. Khadre, T. H. Shellhammer, and A. E. Yousef. 2002. Immunoassay Method for Quantitative Determination of Nisin in Solution and on Polymeric Films. *Lett. Appl. Microbiol.* 34:199–204.

Lyon, W. J., J. K. Sethi, and B. A. Glatz. 1993. Inhibition of Psychrotrophic Organisms by Propionicin PLG-1, a Bacteriocin Produced by *Propionibacterium thoenii*. *J. Dairy Sci.* 76:1506–1513.

Marciset, O., M. C. Jeronimus-Stratingh, B. Mollet, and B. Poolman. 1997. Thermophilin 13, a Nontypical Antilisterial Poration Complex Bacteriocin, That Functions without a Receptor. *J. Biol. Chem.* 272:14277–14284.

Meghrous, J., C. Lacroix, and R. E. Simard. 1999. The Effects on Vegetative Cells and Spores of Three Bacteriocins from Lactic Acid Bacteria. *Food Microbiol.* 16:105–114.

Spelhaug, S. R., and S. K. Harlander. 1989. Inhibition of Foodborne Bacterial Pathogens by Bacteriocins from *Lactococcus lactis* and *Pediococcus pentosaceus*. *J. Food Prot.* 52:856–862.

CHAPTER 11

LACTIC ACID FERMENTATION AND BACTERIOCIN PRODUCTION

BATCH FERMENTATION; GROWTH KINETICS; BACTERIOCIN BIOASSAY

INTRODUCTION

Successful food and industrial fermentations require a good understanding of microbiology, biochemistry, engineering, and other avenues of knowledge. This chapter introduces the subject of fermentation with emphasis on microbiology and food applications. A small-scale fermentation using lactic acid bacteria (LAB), will be demonstrated. Progress of the fermentation will be assessed in terms of microbial growth and product formation.

Fermentors

Although some commercial food fermentations are done in simple apparatus, specialized equipment is needed for the proper control of the process. Fermentations can be carried out in a batch or a continuous mode. Controlled batch fermentations are done in "fermentors" while continuous fermentations require an elaborate setup, called a "bioreactor." The fermentor is an apparatus in which the mixture of substrates and microorganisms is incubated under optimum conditions for the production of desired fermentation end products (Fig. 11.1). Fermentor vessels vary in capacity from a few hundred milliliters to thousands of liters. Factors that can be controlled in the fermentor include temperature, agitation, aeration, and pH. Fermentation can be controlled by automatic or semiautomatic means. Cell density, product formation, or decomposition of media components may be monitored, depending on the goal of the fermentation process.

Food Microbiology By Ahmed E. Yousef and Carolyn Carlstrom
ISBN 0-471-39105-0 Copyright © 2003 by John Wiley & Sons, Inc.

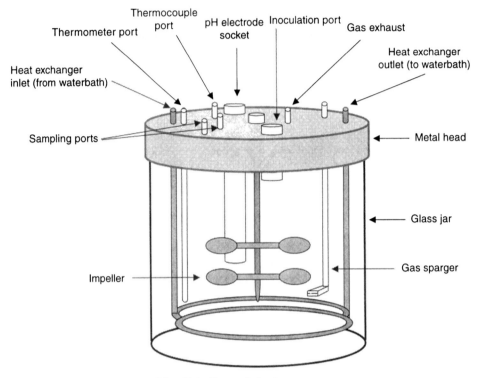

Fig. 11.1. Parts of the fermentor.

Fermentation Monitoring

Experimental bench-top fermentations may provide valuable knowledge about the inoculated microorganism, such as the optimum conditions for rapid growth, consumption of a substrate, or production of a valuable metabolite. To assess the progress in achieving these goals, changes in cell density, consumption of a substrate, or release of a metabolite should be monitored during the course of fermentation. Manual monitoring of fermentation entails taking samples of the fermentation mixture (fermentate) at short intervals (e.g., every hour) and analyzing these samples promptly. Since the fermentation may last for days, manual monitoring becomes a tedious task. Instrumental monitoring of the fermentation is a desirable alternative, but cost of the monitoring devices is generally high. Additionally, there are limitations for measurements taken by instruments, compared to those measured manually. Cell density, for example, may be correlated with medium turbidity that is measured by a spectrophotometer with a flow cell that continuously receives samples of the fermenting mixture (Fig. 11.2). The spectrophotometric measurements (i.e., absorbance units) are then sent to a computer for later retrieval and analysis. Unfortunately, absorbance is only a crude measure of cell density, and results are not equally meaningful at all phases of growth. On the contrary, samples collected manually and plated on suitable media produce counts that reliably demonstrate changes in cell population during all phases of growth. Conversion of carbohydrates into lactic acid during the fermentation may be monitored by meas-

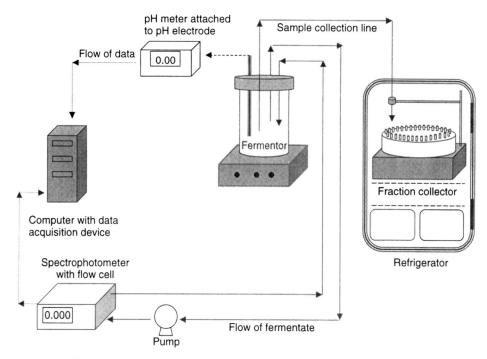

Fig. 11.2. Fermentation monitoring system used in this experiment.

uring and recording the pH of the fermenting medium automatically. However, pH readings cannot be converted into lactic acid concentrations. Metabolites other than acid may require multiple steps of extraction and purification before the concentration is measured. Such metabolites may be monitored only by manual sampling and analysis. Monitoring bacteriocin production during fermentation, for example, requires sample manipulations that may not be easily automated.

Bacteriocin Bioassay

Direct quantitative methods for routine measurements of bacteriocin concentrations are still in the developmental stage. Quantities of bacteriocin released during the fermentation may be measured indirectly by determining the antimicrobial activity of the fermenting medium. Most methods of measuring bacteriocin activity rely on using a sensitive bacterium (indicator) and determining the degree of inactivation or inhibition of this indicator by the medium containing the bacteriocin. These methods are described as "bioassays" because biological indicators are used in measuring the activity. When a reference standard of the bacteriocin of interest is available, the bioassay may be used to measure bacteriocin concentration with a reasonable accuracy. Since pure nisin and preparations with known nisin concentration are commercially available, determining the concentration of this bacteriocin in fermented media is possible. Unfortunately, most bacteriocins are not available commercially in pure forms. In fact, only a limited number of bacteriocins have been purified and chemically defined. Bioassays, in these cases, determine bac-

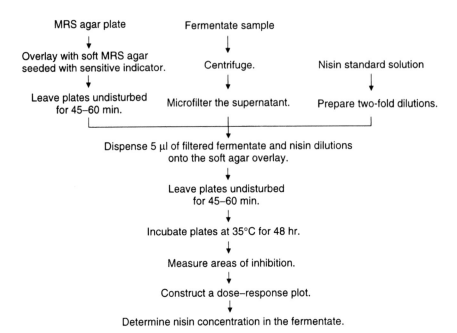

Fig. 11.3. Bacteriocin bioassay in *Lactococcus lactis* fermentate using nisin standard solution as a reference.

teriocin activity in relative units, that is, in arbitrary units. Examples of bioassays with or without reference standard are presented:

1. *Bioassay with Bacteriocin Reference Standard* Dilutions of the bacteriocin are prepared and a small volume (e.g., 5 µl) of each dilution is spotted onto a layer of soft agar medium, seeded with the indicator bacterium (Figs. 11.3 and 11.4). After incubating the agar plates, circular areas of no growth (i.e., clear agar) appear where the solutions with antimicrobial activity are spotted (Fig. 11.5). A dose–response plot is constructed to reveal the correlation between bacteriocin concentration, that is, the dose and area of inhibition observed (the response). It is commonly noticed that the diameter of the inhibition area is proportional to the log bacteriocin concentration. The diameter of the inhibition area produced by the fermentate is matched with a similar point on the standard dose–response plot and the corresponding bacteriocin concentration is determined.

2. *Bioassay without Bacteriocin Standard* Critical dilution assay is a popular method for measuring relative bacteriocin activity when no reference preparation is available. Conceptually, this method is similar to the "minimum inhibitory concentration" technique for measuring antibiotic potency. The critical dilution method involves diluting the culture or fermentate to be tested (normally a twofold dilution series) and spotting small volumes (e.g., 5 µl) of these dilutions onto a layer of soft agar seeded with the indicator bacterium. After incubation of agar plates, areas of inhibition appear where the dilutions with antimicrobial activity are spotted. The highest dilution factor generating an area of inhibition indicates the strength of the antimicrobial action. The bacteriocin activity is, therefore, proportional to the

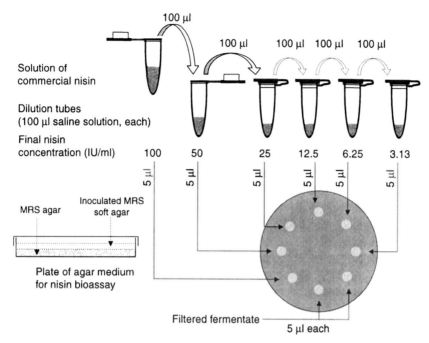

Fig. 11.4. Preparations for nisin bioassay in the product of batch fermentation (fermentate).

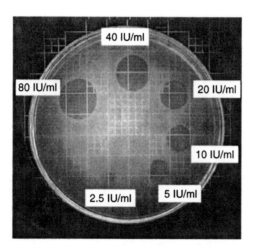

Fig. 11.5. Inhibition areas resulting from spotting dilutions of nisin stock solution onto a soft MRS agar seeded with *Lactobacillus leichmannii* ATCC 4797. (Courtesy of Patsy Leung.)

reciprocal of the highest dilution factor producing a detectable area of inhibition (DF_i). Using this information, a relative bacteriocin activity value, known as activity arbitrary units (AU), is calculated as follows:

$$\frac{AU}{ml\ culture} = \frac{1}{DF_i} \frac{1000}{volume\ spotted\ in\ \mu l}$$

when 5-µl portions are spotted on the agar seeded with the indicator, AU/ml = $200/DF_i$. For example, if a fermentate that was diluted to $1/2^1$–$1/2^8$ produced areas of inhibition (i.e., clear circles) at all dilutions except $1/2^8$, then AU/ml = $200/(1/2^7)$ = 25,600. The AU value varies with the indicator microorganism and test conditions.

OBJECTIVES

1. Learn about controlled batch fermentations.
2. Construct a growth curve and identify phases of microbial growth.
3. Monitor the accumulation of beneficial fermentation products (i.e., lactic acid and nisin).

PROCEDURE OVERVIEW

In this exercise, a controlled batch fermentation will be carried out in a 3-liter fermentor (Fig. 11.1) using a nisin-producing lactic acid bacterium (*Lactococcus lactis* subsp. *lactis* ATCC 11454). The total fermentation time is 24 hr. Changes in cell density, pH, and bacteriocin activity will be monitored since these are the main contributing factors to food preservation by LAB. The laboratory exercise can be limited to growth and acid monitoring. However, if a bacteriocin-producing culture is used, it will be advantageous to run the complete exercise.

The fermentation will be monitored as shown in Fig. 11.2. A spectrophotometer and a pH meter connected to a computer with a data acquisition device continuously measure cell turbidity and pH during the fermentation. Samples of the fermentate are also collected by a small pump, and equal volumes (e.g., 4–5 ml) at equal time intervals (e.g., 30 min) are delivered in test tubes using a fraction collector. The fraction collector with collected samples is kept in a refrigerator set at 1–4°C during the fermentation and the short-term overnight sample storage. Refrigeration temporarily arrests growth and metabolite production by the culture in the collected samples.

The suggested scheme results in 48 samples collected during the 24 hr of incubation. If all samples cannot be assigned to students, selected samples representing all phases of growth are analyzed. Since little or no bacteriocin is produced during the first few hours of fermentation, samples from this period are analyzed less frequently for bacteriocin activity than those from subsequent fermentation periods.

In the first laboratory period, students will inspect the fermentor and fermentation monitoring devices. A culture of *L. lactis* is inoculated in a deMan, Rogosa, and Sharpe (MRS) broth medium fermentor. Changes in absorbance and pH during fermentation will be monitored and recorded electronically. Samples of the fermenting medium will be collected automatically at equal time intervals and refrigerated immediately as described earlier.

In Period 2, each student is assigned a time-labeled fermentate sample (e.g., a tube containing fermentate collected after 3–$3\frac{1}{2}$ hr of inoculation). Half of the sample is transferred to a sterile vial and frozen at −20°C for bacteriocin bioassay during the subsequent laboratory period. The remainder of the sample is analyzed for total plate count as explained in Chapter 2 of this manual with these modifications: (a) dilutions are made in sterile microcentrifuge tubes using a total volume of 1000 µl/tube and (b) the MRS agar medium is used for counting *L. lactis* in the fermentate. Data on pH and cell density (measured as A_{600nm}) that were collected automatically during the fermentation are shared.

During the third laboratory period, students will inspect the incubated MRS agar plates that were prepared in the previous period. Colonies are counted and total count in the fermentate sample is determined following the counting rules (see Chapter 1). *Lactococcus lactis* counts in all time-labeled samples are pooled (class data) and plotted against fermentation time to generate a growth curve. These data are used to calculate generation time and other growth parameters.

Bioassay for bacteriocin in the fermentate is also carried out during the third laboratory period. Each student obtains the frozen fermentate sample that was prepared in the previous laboratory period. These samples are thawed and analyzed for nisin activity as explained in the introduction of this chapter (bioassay with reference standard). Nisin is active against several foodborne pathogens, but a nonpathogenic, nisin-sensitive indicator bacterium, *Lactobacillus leichmannii* ATCC 4797, will be used as an indicator in this assay (Table IV.2).

In Period 4, agar plates for bacteriocin bioassay are inspected and nisin concentration in the fermentate is determined. The presence of areas of inhibition indicates antimicrobial activity in the fermentate. Diameters of areas of inhibition resulting from the spots of nisin solutions are measured and plotted against $\log_{10}$ nisin concentration to create a dose–response plot. The diameter of the inhibition area resulting from spotting the fermentate is located on the dose–response plot and the corresponding nisin concentration is determined. Nisin concentrations in all the time-labeled samples are pooled and plotted against fermentation time.

MEDIA

Lactobacilli MRS Broth

This medium will be referred to as "MRS broth." This medium supports growth of most LAB, particularly the slow-growing strains of lactobacilli. The medium is nutritionally rich since it contains peptone, beef extract, yeast extract, and glucose. Fastidious microorganisms such as LAB grow luxuriously on this medium. The medium is somewhat selective for lactobacilli since it contains Tween 80 and sodium acetate.

Lactobacilli MRS Agar

The medium is prepared from MRS broth by addition of agar to a final concentration of 1.5%.

Soft MRS Agar

The medium is prepared from MRS broth by addition of agar to a final concentration of 0.75%.

ORGANIZATION

Students will work individually. Each student is assigned a time-labeled fermentate sample. Half of the sample is analyzed for total count during the second laboratory period and the remainder is used for bacteriocin assay during subsequent periods.

REFERENCES

Gagliano, V. J., and R. D. Hinsdill. 1970. Characterization of a *Staphylococcus aureus* Bacteriocin. *J. Bacteriol.* 104:117–125.

Klaenhammer, T. R. 1993. Genetics of Bacteriocins Produced by Lactic Acid Bacteria. *FEMS Microbiol.* 34:199–204.

Leung, P. P., M. A. Khadre, T. H. Shellhammer, and A. E. Yousef. 2002. Immunoassay Method for Quantitative Determination of Nisin in Solution and on Polymeric Films. *Lett. Appl. Microbiol.* 34:199–204.

Liao, C.-C., A. E. Yousef, E. R. Richter, and G. W. Chism. 1993. *Pediococcus acidilactici* PO2 Bacteriocin Production in Whey Permeate and Inhibition of *Listeria monocytogenes* in Foods. *J. Food Sci.* 58:430–434.

Period 1 Fermentation

MATERIALS AND EQUIPMENT

Class Shared

- Bench-top fermentor display (dismantled)
- Bench-top fermentor setup
- *Lactococcus lactis* subsp. *lactis* ATCC 11454 (16-hr culture)

PROCEDURE

1. Inspect the demonstrated bench-top fermentor. Observe the important components, including the agitation mechanism, sparger, temperature and pH-measuring devices, and inoculation and sampling ports. Compare the fermentor displayed with that illustrated in Fig. 11.1.
2. Observe the assembled fermentor, including the temperature and pH monitoring and recording devices, and the sample collection scheme. Compare the actual setup with that illustrated in Fig. 11.2. Notice that the fermentor jar with its contents of MRS broth and the pH probe have been autoclaved as one unit. Sampling devices are autoclaved separately and connected to the fermentor jar using aseptic technique.
3. Observe the inoculation of the fermentation medium as done by the laboratory instructor. *Lactococcus lactis* subsp. *lactis* ATCC 11454 is inoculated at the 0.001% level into 2 liters of MRS broth medium in the fermentor.
4. The fermentation vessel is held at 30°C, purged with N_2, and stirred at approximately 200 rpm. The batch fermentation is terminated after 24 hr of inoculation.
5. Absorbance at 600 nm (A_{600}) and pH of fermented medium are measured continuously using a spectrophotometer and pH meter, respectively, and data are recorded electronically.
6. Samples are withdrawn continuously and refrigerated immediately at 1–4°C. It is recommended that samples are taken at 30-min intervals and 4 ml is collected in each tube. After the fermentation is complete, label the samples sequentially and return the tubes to the refrigerator promptly. Prepare a table that associates the tube number with the fermentation time. This step may be carried out by the laboratory instructor.

    [*Example*: If a sample was collected every 30 min, the first tube receives a sample collected between 0 and 30 min of fermentation. This tube is labeled (#1) and average fermentation time for the sample in this tube is (0 + 30)/2, or 15 min.]
7. The refrigerated fermentate is analyzed during subsequent laboratory periods for total count and bacteriocin activity.

Period 2 Total Plate Count

MATERIALS AND EQUIPMENT

Per Student

- Time-labeled fermentate sample
- A sterile 4- or 5-ml freezer vial
- Micropipetter (0–1000 µl)
- Test tube containing 10 ml sterile diluent (0.85% saline solution)
- Six plates containing MRS agar
- Sterile microcentrifuge tubes (~1.5 ml capacity)

    (*Note:* These tubes will be used to prepare sample dilutions. The number needed depends on the stage of growth at the time of collecting the fermentate sample. This number should be estimated by the students before running the exercise. These tubes have been autoclaved in an aluminum foil–sealed beaker. The tubes should be handled carefully, and proper aseptic technique applied.)

Class Shared

- Incubator, set at 30°C

PROCEDURE

1. Record the tube number and the corresponding sampling time of the chosen fermentate sample:

    Tube #:_____   Sampling time:_____ hr

2. Dispense 2 ml of the fermentate into the freezer vial and freeze at –20°C immediately. This sample is used in the subsequent laboratory period for bacteriocin bioassay.
3. Pick a number of microcentrifuge tubes and MRS agar plates based on the estimation of the number of dilutions needed for the assigned fermentate sample.
4. Label the microcentrifuge tubes with appropriate dilution factors.
5. Dispense 900 µl diluent into each of the microcentrifuge tubes.
6. Vortex the fermentate sample. Transfer 100 µl of fermentate into the microcentrifuge tube receiving the 10^{-1} dilution. Vortex the diluted sample.
7. Prepare the remaining sample dilutions, as needed.
8. Select three dilutions for plating. (*Note:* Normally, the highest three dilutions are plated.) Since plating is done in duplicate, six agar plates are needed. Label these agar plates accordingly.
9. Dispense 100 µl of appropriate dilution onto the corresponding MRS agar plates. Spread the inoculum on agar surface with a sterile spreader.
10. Incubate the plates at 30°C for 48 hr.

Period 3 Total Count (*Continued*)

MATERIALS AND EQUIPMENT

Per Student

- Incubated MRS agar plates (prepared in the previous laboratory period)

Class Shared

- Colony counters

PROCEDURE

1. Inspect the incubated MRS agar plates and determine those containing colonies in the countable range. (See Chapter 1 for counting rules.)
2. Count colonies and determine CFU/ml fermentate.

   Sample number: _____
   Sampling time: _____ hr
   Dilutions plated: _____
   Dilution with countable plate: _____
   Counts of colonies on countable plates: _____
   CFU/ml: _____

3. Record these results in Table 11.1; include data of classmates in this table (provided by laboratory instructor.)
4. Obtain a copy of all data points gathered during the fermentation process. Use linear-scale graphing paper (see Appendix B) to plot these data on a growth curve. The Y axis represents log CFU/ml and the X axis represents sampling time during the fermentation. If log-linear graphing paper is used, then the Y axis should represent CFU/ml (i.e., the untransformed data).
5. Identify the approximate time for the beginning and the end of each of these growth phases: lag, exponential, and stationary (see Part IV for details).
6. Using an appropriate segment of the exponential growth phase, determine the generation time and specific growth rate for *L. lactis* under these fermentation conditions.
7. Determine the maximum growth (CFU/ml) achieved during this fermentation process.
8. Plot the automatically gathered data (A_{600nm} and pH) against fermentation time using graphing paper with linear-linear axes. Use a double-Y plot where the left Y axis represents A_{600nm}, the right Y axis represents the pH, and the X axis represents fermentation time.

TABLE 11.1. Total Plate Count (CFU[a]/ml) in Different Fractions of Fermentate as Reported by Classmates

Time (hr)	Count (CFU/ml)	Time (hr)	Count (CFU/ml)

[a] Colony-forming units.

Period 3 (*Continued*) Bacteriocin Bioassay
(Nisin Used as a Reference Standard)

MATERIALS AND EQUIPMENT

Per Student

- Frozen time-labeled fermentate sample
- Seven sterile microcentrifuge tubes
- Sterile 2- or 3-ml syringe
- Sterile 0.2-µm filter
- Micropipetter (0–20 µl)
- Micropipetter (0–1000 µl)
- Tube containing 2 ml sterile diluent (0.85% saline solution)
- Overnight culture of *Lactobacillus leichmannii* ATCC 4797
- A 9-ml tube of molten MRS soft agar (contains 0.75% agar) (*Note*: Remove from the waterbath only when ready to use.)
- Petri plate containing MRS agar
- Microcentrifuge tube containing 200 µl nisin solution standard (100 IU/ml) [*Note*: One microgram of pure nisin equals 40 international units (IU).]

Class Shared

- Centrifuge (microcentrifuge)
- Waterbath, set at ~50°C
- Incubator, set at 35°C

PROCEDURE

Preparing Soft-Agar Overlay

1. Prepare the mixture of nisin-sensitive indicator and soft agar as follows:
   a. Transfer 10 µl of an overnight culture of the sensitive bacterial strain (*L. leichmannii* ATCC 4797) into 9 ml of molten MRS soft agar that has been precooled in a waterbath to about 50°C.
   b. Mix gently by rolling between hands. (*Note*: Vigorous shaking incorporates air bubbles in the agar, which when poured produces a rough surface for bacteriocin spotting.)
2. Pour the contents of the inoculated soft-agar tube into a previously prepared plate of regular MRS agar. Do not agitate or swirl the plate.
3. Let the soft-agar overlay solidify by leaving the covered plates on the bench undisturbed. Although solidification of the agar overlay takes only a few minutes at room temperature (~22°C), it is advisable to leave these plates for 45–60 min before spotting the filtrates; this minimizes diffusion of spots and results in more round inhibition areas.

Preparing Nisin Standard Solution

Prepare a series of twofold dilutions of the standard nisin solution (100 IU/ml). Five dilution tubes are required (Fig. 11.4). Notice that in a twofold dilution, the volume of diluent in the tube equals the volume being transferred.

1. Place 100 μl of the diluent into five separate sterile microcentrifuge tubes. Label the tubes with these calculated nisin concentration: 50, 25, 12.5, 6.25, and 3.13 IU/ml.
2. To the 50-IU tube, add 100 μl of the original standard solution (100 IU/ml).
3. Vortex the 50-IU tube, then transfer 100 μl of the tube contents to the 25-IU tube.
4. Repeat until completing the 3.13-IU tube.

Preparing Fermentate

1. Record the tube number and the corresponding sampling time of the chosen fermentate sample.

    Tube #:_____   Sampling time:_____ hr

2. Thaw the frozen fermentate sample in a beaker containing warm water and mix the contents by vortexing.
3. Transfer, aseptically, 1 ml of the fermentate to a sterile microcentrifuge tube.
4. Centrifuge the microcentrifuge tube at 3000 rpm for 2 min.
5. Transfer the supernatant (the cell-free clear liquid) into a 2-ml (or 3-ml) syringe. Filter the sample through the 0.2-μm filter. Collect the filtered supernatant into a sterile microcentrifuge tube (Fig. 11.6). This procedure can be done in this sequence:
    a. Open the microfilter package, exposing only the side of the filter to be fitted to the syringe barrel.
    b. Remove the syringe plunger and any tip-guard or needle and screw the filter onto the syringe barrel.
    c. Remove the remainder of the microfilter package and place the sterile end of the microfilter onto the sterile microcentrifuge tube (on a microcentrifuge tube rack).
    d. Pour the supernatant of culture into the syringe barrel.
    e. Attach syringe plunger and push it gently to drive the supernatant through the filter. Filtration may be stopped when at least 150–250 μl of sterile filtrate is collected in the microcentrifuge tube.

Spotting

1. Label, appropriately, the bottom of the plate containing the MRS agar and overlay. Mark evenly the locations to be spotted with the nisin solution or the fermentate (Fig. 11.4).

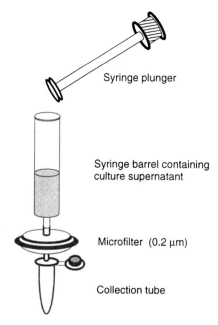

Fig. 11.6. Microfiltering supernatant of bacteriocin-producing culture.

2. Apply 5 µl (using the 20-µl micropipetter) of the filtered fermentate onto the surface of soft-agar overlay. Keep the pipetter vertical to produce as round and even a spot as possible.
3. Repeat the previous step so that a duplicate spot of the fermentate is produced.
4. Repeat step 2 using the nisin standard solutions. Evenly space the spots on the plate.
5. Keep the plates undisturbed until all spots are absorbed into the agar (30–60 min).
6. Incubate the plate at 35°C for 48 hr.

Period 4 Bacteriocin Bioassay (*Continued*)

MATERIALS AND EQUIPMENT

Per Student

- Incubated MRS agar plate, prepared during the previous laboratory period
- A 10-cm ruler (transparent plastic rulers with 0.5-mm divisions are ideal)

Class Shared

- Colony counters

PROCEDURE

1. Inspect the incubated MRS agar plate using a colony counter and observe areas of inhibition. Spotted nisin solutions should produce distinctly clear areas of inhibition.
2. Determine if the spots from the filtered fermentate produced areas of inhibtion.
3. Measure diameters of all inhibition areas using the ruler. Record these results in Table 11.2.
4. Using graphing paper with linear axes, plot $\log_{10}$ nisin concentration (X axis) against diameter of inhibition area (Y axis) for the nisin standard solutions. Label this plot as "dose–response plot."
5. Calculate the average diameter of the inhibition areas for the fermentate. Using the dose–response plot, determine the $\log_{10}$ concentration of nisin that corresponds to this diameter. The anti-log of this value is the estimated nisin concentration in the fermentate sample.
6. Report nisin concentration to the instructor. Copy from the instructor's table the classmates' data into Table 11.3.
7. Prepare a plot depicting changes in nisin concentration (IU/ml) during the fermentation using data from Table 11.3. Plot IU/ml on the Y axis and fermentation time on the X axis. Use graphing paper with linear axes (see Appendix B).
8. Compare this plot with those prepared in the previous laboratory period. Describe the relation between growth phase and bacteriocin production.

TABLE 11.2. Diameters of Inhibition Areas Resulting from Spotting 5 µl of Nisin Standard Solutions and Filtered Fermentate (Sample Number _____)

Nisin Standard		Fermentate	
Concentration (IU/ml)[a]	Diameter (mm)	Spot Number	Diameter (mm)
100		1	
50		2	
25		Average diameter (fermentate) =	
12.5			
6.25			
3.13			

[a] International units per milliliter of fermentate.

TABLE 11.3. Nisin Concentrations (IU/ml)[a] in Different Fractions of Fermentate as Reported by Classmates

Time (hr)	Concentration (IU/ml)	Time (hr)	Concentration (IU/ml)

[a] International units per milliliter of fermentate.

PROBLEMS

1. In this exercise, cell density, pH, and bacteriocin release were monitored during the fermentation. List at least two additional measurements that may vary during the fermentation and indicate how these measurements can be taken.

2. Define *bacteriocin*. How is a bacteriocin different from an antibiotic?

3. Summarize the classification of bacteriocins produced by lactic acid bacteria as described by Klaenhammer, 1993 (see the References).

4. What is the function of each of the following devices in the fermentation process:

    (a) Computer
    (b) Thermocouple
    (c) Fraction collector

5. Use the information provided in the laboratory exercise about inoculum size and properties of the culture to estimate the count of *L. lactis* in the fermentor immediately after inoculation. Compare this estimate with the result obtained from analyzing the first fermentate sample for total plate count.

6. Plot the automatically gathered data (A_{600nm} and pH) against fermentation time, as instructed earlier. Label the axes and provide a suitable title. Mark this plot as Fig. 11.A.

7. What are the advantages and disadvantages of monitoring cell density during the fermentation by reading the absorbance?

8. Plot the pooled class data on total plate count to construct a growth curve as described earlier. Label the axes and provide a suitable title. Mark this plot as Fig. 11.B.

    (a) Mark on the graph the lag, exponential, and stationary phases.
    (b) Determine the minimum generation time and maximum specific growth rate. Mark the segment of the growth curve that was used to generate these values.
    (c) Determine the maximum growth (CFU/ml) achieved during this fermentation process.
    (d) Compare Figs. 11.A and 11.B with regard to suitability for defining phases of growth.

9. Use the pooled class data to draw the bacteriocin concentration curve as described earlier. Label the axes and provide a suitable title. Mark this plot as Fig. 11.C.

10. Use Figs. 11.B and 11.C to describe the relation between growth phase and bacteriocin production.

11. What are the advantages and disadvantages of using AU/ml as a measure of bacteriocin production?

APPENDIX A

LABORATORY EXERCISE REPORT

Laboratory reports may be prepared for some or all of the laboratory exercises in this manual. The reports make the laboratory training useful and meaningful. A *laboratory report* should be concise and contain only information relevant to the exercise. In some exercises, a worksheet instead of a report may be prepared. *Laboratory worksheets* are mainly made of tables filled in while students are running the exercises, with brief comments on the gathered data. Worksheet structure, however, may be designed to train students how to prepare a complete laboratory report. Therefore, questions pertinent to different sections of a typical laboratory report may be included. Using this expanded definition of a worksheet, students are required to answer these questions and type their answers.

After the first report is graded, students may be allowed to rewrite the report to address the problems they encountered earlier. Students who choose to rewrite the report can earn up to 50% of the points they lost in the first round of grading. If a laboratory rewrite is prepared, students should submit both the original and the rewritten reports within a week after receiving the graded reports.

LABORATORY REPORT OUTLINE

The laboratory report should include the following sections:

1. Cover page
2. Introduction
3. Objectives
4. Methods
5. Results
6. Discussion
7. References

Students may be asked to answer specific questions and include their answers in relevant sections of the report.

Cover Page

The cover page should include the following:

Laboratory exercise title: (This is the chapter title as shown in laboratory manual.)
Name: (**Class partner's name:** _____)
Date due:
Descriptive title:

Introduction

The introduction may include information pertaining to the morphological and biochemical properties of the microorganism(s) tested. The introduction should include brief comments on the importance of the organism in the food industry. Additionally, the following questions may be addressed in the introduction:

Does the food analyzed support growth or survival of the microorganism of interest?

If growth is not supported, does the food tested inhibit or inactivate such a microorganism?

Under which conditions would the microorganism proliferate in the food analyzed?

If there are standards setting an unacceptable level of this microorganism in food, what are these standards?

Objectives

The objectives should be brief and explain in one sentence why the laboratory was performed. Although a set objectives has been included with each laboratory exercise in this manual, students should rewrite these in their own words.

Methods

Since detailed methodologies are included already in the manual, the Methods section of the laboratory exercise report should be very short. Cite the manual, as the source of the procedure, in this section of the report, indicating which method was followed. Additionally, this section covers the following:

- *Information about the sample used in the analysis.* These may include (a) name, (b) method of packaging (e.g., canned, vacuum packaged), (c) storage conditions (e.g., refrigerated, frozen, shelf stable), (d) package size, and (e) sell-by date.
- *Discrepancies between the method of the manual and actual procedures performed.*

Results

This section should include all data presented in organized formats such as tables and graphs. Assign numbers and titles to all tables and graphs. Include footnotes with tables to explain units and abbreviations (e.g., CFU/ml) or to explain irregularities. Label all axes on graphs and assign units to all numbers. When results require calculations, show one example only of each type of calculation and include units.

The Results section should include *individual data and group data*, that is, data gathered by the student only and by the group (class partners). Students may occasionally be asked to comment on *class data*, that is, data collected by all individuals or all groups in the class. Class data may be listed in a table containing individual or group data or reported as an average.

Discussion

In this section, results are explained. Therefore, reference to the tables and numbers in the Results section should be emphasized. If the food is judged to be of poor microbiological quality, explain what could have been the cause of the contamination (e.g., cross-contamination or underprocessed). Make plausible assumptions about the microbiological quality and safety of food tested. Whenever appropriate, the student may speculate whether a customer (e.g., a food retailer) would accept or reject a batch of food from which samples gave similar results.

References

Students should cite facts or data from sources outside of the manual or class notes. Two references at least, including books, research papers, review articles and websites, should be consulted and cited. For citations and listing of references, see the example laboratory report below. The guidelines of the American Society for Microbiology (ASM) for reference citing are recommended.

TABLES AND FIGURES

Tables and figures are essential components of laboratory reports. Well-structured tables and figures improve the readability of the report. Each table or figure should be informative and self-explanatory. Some aspects of figure construction have been addressed in Part IV of this manual.

TABLE A.1. Counts of Coliforms[a] in Ground Meat from Different Sources

Meat Sources	Count/g[b]
Mom & Pop shop	5.0×10^3
Giant Supermarket	1.0×10^3
Meat teaching facility	$<1.0 \times 10^2$

[a] Count was determined using violet red bile agar.
[b] Results are presented as colony forming unit (CFU)/g.

The table has a title, headings, and contents (see Table A.1). The table title refers to the (a) data being presented in the table's contents (e.g., counts of coliforms) and (b) variables tested by these data (e.g., source of meat). The heading of a column should always describe the contents of this column. Footnotes define terms or explain conditions.

REPORT EXAMPLE

The following is an example of a laboratory report. Data in this example are imaginary but adequate for demonstrating the structure of a report. Note that a laboratory report is typed and lines are double spaced.

Laboratory exercise title: *Campylobacter* in Selected Foods
Name: John Smith   {Class partner's name: Smith Jones}
Date due: xx/xx/xx
Descriptive title: Enumeration and isolation of *Campylobacter* spp. in raw milk using different selective media.

INTRODUCTION

Campylobacter spp. are gram-negative non-spore-forming bacteria with S- or spiral-shaped cells. They are microaerophilic to anaerobic, oxidase positive, catalase positive, and highly motile. These organisms can grow at 42°C but not at 25°C. This ability to grow at 42°C and not at 25°C is used as a means of identifying an isolate as *Campylobacter*. The cell size of *Campylobacter* ranges from 0.2 to 0.5 µm wide and 1.5–5 µm long, and thus the microorganism can be separated from most other gram-negative bacteria by using a 0.65-µm filter (2).

Most *Campylobacter* strains are associated with acute gastroenteritis in humans. These organisms are pathogens of many animals. *Campylobacter* spp. have been found in fecal contents of slaughtered sheep, cattle, and other animals. Therefore, meat, milk, and poultry could be contaminated with this organism, and ingestion of infected or contaminated food could cause a serious infection (2). The hazard of *Campylobacter* in food can be minimized by preventing cross-contamination. Consumption of undercooked or unpasteurized foods of animal origin should be avoided.

Since foods of animal origin are likely to be contaminated with *Campylobacter* spp. (1), milk was tested in this laboratory exercise. In fact, raw milk is the most common food-associated *Campylobacter* infection. Milk is a very nutritious medium, but conditions of production and storage probably do not support the growth of this pathogen. *Campylobacter* does not grow at refrigeration temperatures similar to those used for milk storage. Milk probably contains sufficient dissolved oxygen to prevent the growth of this microaerophilic pathogen. *Campylobacter* spp., however, may survive in raw milk, and cross-contamination of the pasteurized product is likely to occur. Different media are currently available for detection of *Campylobacter* spp (1). Some of these media were tested in this laboratory exercise to recover the pathogen from milk.

OBJECTIVE

Detect *Campylobacter* in raw milk and compare isolation media and techniques.

METHODS

Raw whole milk was obtained from the university farm in a 5-gallon milk can. The milk was produced 24 hr before running the laboratory exercise. It was kept in the laboratory walk-in cooler; temperature was 6.0°C at the time of sampling. The procedures for the isolation and detection of *Campylobacter* were those outlined in the laboratory manual with two deviations. During centrifugation of milk to concentrate contaminant microorganisms, a 250-ml sample was centrifuged at 15,000 g for 25 min instead of 16,300 g; 15,000 g is the maximum speed for the centrifuge that was available in the laboratory. For plating, Preston agar was substituted for Butzler's agar.

RESULTS

TABLE 1. Detection of *Campylobacter* in Cell Concentrate from Raw Milk by Direct Plating and Plating after Preenrichment[a]

Medium	Direct Plating	Preenrichment
Preston agar	−/−[b]	−/−
Campy–Line agar	−/+[c]	+/+
Skirrow's agar	−/−	+/+

[a] Results are for duplicate plates.
[b] Not detected.
[c] Detected.

TABLE 2. Enumeration of *Campylobacter* in Cell Concentrate from Raw Milk Using Campy–Line Agar

	Colonies Counted	
Dilution[a]	Plate 1	Plate 2
10^0	1	3
10^{-1}	0	0
10^{-2}	0	0

[a] One milliliter of each dilution was used and pour-plating technique was followed.

Sample calculation:

$$\text{CFU/ml milk concentrate} = \frac{\text{average count of most countable plates}}{\text{volume} \times \text{dilution factor}}$$

$$= \frac{(1+3)/2}{1 \times 10^0}$$

$$= 2.0 \times 10^0 \text{ CFU/ml (est.)}$$

DISCUSSION

Campylobacter spp. were detected in a cell concentrate from raw milk. According to results in Table 1, Campy–Line agar was better than Preston agar and Skirrow's agar media in recovering *Campylobacter* spp. from the sample by direct plating. Preenriching the concentrate before plating, however, improved the overall recovery of *Campylobacter* spp. from the sample. While direct plating resulted in only one positive plate (of six, total), preenrichment before plating produced four positive plates. No *Campylobacter* spp. were recovered from Preston agar medium by direct or indirect plating. It is apparent that Campy–Line agar is superior than the other tested media in detecting *Campylobacter* spp. Additionally, preenrichment of samples before plating improved the recovery of the pathogen from milk. In the latter case, both Campy–Line agar and Skirrow's agar media were equally effective in recovering *Campylobacter* spp. from milk.

Enumeration of *Campylobacter* in cell concentrate from raw milk using Campy–Line agar produced a very small number (2.0×10^0 CFU/ml concentrate, est.) (Table 2). This small count may not be a public health concern because (a) milk does not seem to support the growth of this pathogen, (b) *Campylobacter* spp. do not grow during refrigeration and milk is always stored under refrigeration, and (c) pasteurization of milk should eliminate this small population of the pathogen.

REFERENCES

1. George, H. A., P. S. Hoffman, R. M. Smibert, and N. R. Krieg. 1978. Improved media for growth and aerotolerance of *Campylobacter fetus*. J. Clin. Microbiol. 8: 36–41.
2. Park, C. E., R. M. Smibert, M. J. Blaser, C. Vanderzant, and N. J. Stern. 1984. *Campylobacter* pp. 475–496. *In* M. Speck (ed.), Compendium of methods for the microbiological examination of foods, 2nd ed. American Public Health Association, Washington, DC.

APPENDIX B

MICROBIAL GROWTH KINETICS

MODELING THE EXPONENTIAL PHASE

Specific Growth Rate

Growth rate (dx/dt) is proportional to count (x); t is incubation time:

$$\frac{dx}{dt} \propto x \quad \Rightarrow \quad \frac{dx}{dt} = \mu x \quad \Rightarrow \quad \frac{dx}{x} = \mu \, dt \tag{1}$$

During the exponential growth phase, known the *specific growth rate* μ is constant. Integrating both sides of Eq. (1) over specified limits (e.g., from x_1 to x_2 and t_1 to t_2) yields

$$\ln \frac{x_2}{x_1} = \mu \, \Delta t \tag{2}$$

In general terms,

$$\ln x = \ln x_0 + \mu t$$

which is a linear equation, or

$$x = x_0 e^{\mu t}$$

which is similar to that describing the kinetics of a first-order chemical reaction. From Eq. (2)

$$\mu = \frac{\ln x_2 - \ln x_1}{t_2 - t_1} \tag{3}$$

or

$$\mu = \frac{2.3(\log_{10} x_2 - \log_{10} x_1)}{t_2 - t_1} \tag{4}$$

Generation Time (GT)

If $x_2 = 2x_1$ in Eq. (2), then $\Delta t = GT$:

$$GT = \frac{\ln 2}{\mu} = \frac{0.693}{\mu} \tag{5}$$

Combining Eqs. (3) and (5) yields

$$GT = \frac{0.693(t_2 - t_1)}{\ln x_2 - \ln x_1}$$

Then

$$GT = \frac{0.693(t_2 - t_1)}{2.3(\log x_2 - \log x_1)}$$

or

$$GT = \frac{t_2 - t_1}{3.3(\log x_2 - \log x_1)} \tag{6}$$

PROBLEMS FOR PRACTICE

1. Calculate the GT of a bacterium that grows during the exponential phase at the rate of $1 \log_{10}$ CFU/hr.
2. Growth of a bacterium in a liquid medium was monitored for 13 hr and the following results were obtained:

Time (hr)	Count (CFU/ml)	Time (hr)	Count (CFU/ml)
0	3.0×10^3	7.0	9.0×10^6
1.0	2.0×10^3	8.0	4.0×10^7
2.0	3.0×10^3	9.0	8.0×10^7
3.0	2.0×10^3	10.0	1.5×10^8
4.0	5.0×10^3	11.0	2.0×10^8
5.0	3.0×10^4	12.0	2.5×10^8
6.0	5.0×10^5	13.0	2.5×10^8

a. Draw the growth curve of this bacterium using the accompanying graph paper. (*Hint*: If choosing the linear-linear graph, plot $\log_{10}$ CFU/ml values on the Y axis. On the semi-log graph, plot CFU/ml on the Y axis)
b. Measure the following (show the calculations, when applicable, and the units):

   Lag period
   Maximum specific growth rate
   Minimum generation time
   Maximum growth

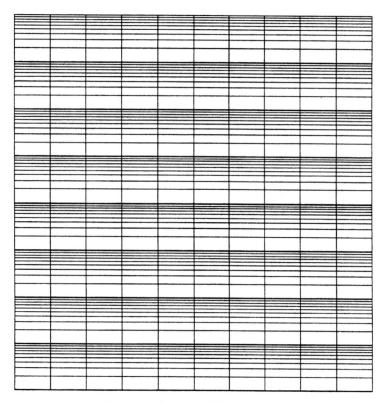

Appendix B.1

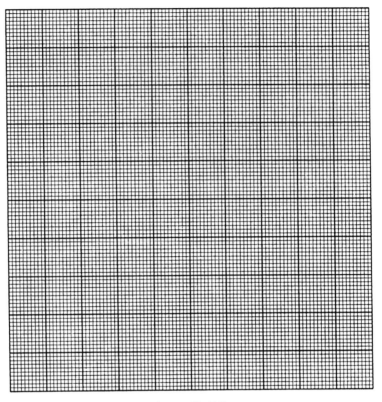

Appendix B.2

APPENDIX C

MICROBIOLOGICAL MEDIA

This list includes microbiological media relevant to the exercises presented in this manual. The composition of these media is based on a formula published in the 11th edition of *Difco Manual*, 1998 (Difco Laboratories, Sparks, MD). All formula ingredients are amounts per liter of distilled water.

BAIRD–PARKER AGAR WITH EGG YOLK/ TELLURITE SUPPLEMENT

Agar	20 g
Beef extract	5
Glycine	12
Lithium chloride	5
Sodium pyruvate	10
Tryptone	10
Yeast extract	1
Egg yolk/tellurite enrichment	50 ml

Final pH: 6.9 at 25°C.

Preparation: Suspend 63 g Baird–Parker agar base in 950 mL distilled or deionized water, heat to boiling, autoclave, cool to 45–50°C, add warm tellurite enrichment, and dispense to plates.

BISMUTH SULFITE (BS) AGAR

Agar	20 g
Beef extract	5
Bismuth sulfite indicator	8

Brilliant green	0.025
Disodium phosphate	4
Ferrous sulfate	0.3
Glucose	5
Peptone	10

Final pH 7.7 at 25°C.

Preparation: Suspend 52 g of BS agar base in 1 liter of distilled or deionized water, heat to boiling, do NOT autoclave, cool to 45–50°C, and dispense to plates.

BRAIN HEART INFUSION (BHI) AGAR

Agar	15 g
Disodium phosphate	2.5
Glucose	2
Infusion from beef heart	250
Infusion from calf brains	200
Proteose peptone	10
Sodium chloride	5

Final pH 7.4 at 25°C.

Preparation: Suspend 52 g of BHI agar in 1 liter distilled water, heat to boiling, autoclave, cool to 45–50°C, and dispense to plates.

BRAIN HEART INFUSION (BHI) BROTH

Same recipe for BHI broth but excludes agar.

BRILLIANT GREEN LACTOSE BILE (BGLB) BROTH

Brilliant green	0.0133 g
Lactose	10
Oxgall	20
Peptone	10

Final pH 7.2 at 25°C.

Preparation: Suspend 40 g of BGLB broth in 1 liter of distilled or deionized water, warm slightly, dispense to tubes containing inverted Durham tubes, and autoclave.

COLUMBIA BLOOD AGAR BASE

| Agar | 15 g |
| Bitone | 10 |

Corn starch	1
Pantone	10
Sodium chloride	5
Tryptic digest of beef heart	3

Final pH 7.3 at 25°C.

Preparation: Suspend 44 g of Columbia blood agar base in 1 liter of distilled or deionized water, heat to boiling, autoclave, cool to 45–50°C, add 5% sterile defibrinated blood at 45–50°C, mix, and dispense to plates.

EOSIN METHYLENE BLUE (EMB) AGAR

Agar	13.5 g
Dipotassium phosphate	2
Eosin Y	0.4
Lactose	5
Methylene blue	0.065
Peptone	10
Sucrose	5

Final pH 7.2 at 25°C.

Preparation: Suspend 36 g of EMB broth in 1 liter of distilled or deionized water, heat to boiling, autoclave, cool to 45–50°C, and dispense to plates.

FRASER BROTH BASE

Acriflavine hydrochloride	0.024 g
Beef extract	5
Esculin	1
Lithium chloride	3
Nalidixic acid	0.02
Potassium phosphate, monobasic	1.35
Sodium chloride	20
Sodium phosphate, dibasic	9.6
Tryptose	10
Yeast extract	5
Fraser broth supplement	10 ml

Final pH 7.2 at 25°C.

Preparation: Suspend 55 g of Fraser broth base in 1 liter of distilled or deionized water, heat to boiling, autoclave, cool to 45–50°C, add Fraser broth supplement, mix, and dispense.

Fraser broth supplement
(per 10-ml vial)

Ferric ammonium citrate	0.5

GRAM-NEGATIVE (GN) BROTH

Dipotassium phosphate	4 g
D-Mannitol	2
Glucose	1
Monopotassium phosphate	1.5
Sodium chloride	5
Sodium citrate	5
Sodium desoxycholate	0.5
Tryptose	20

Final pH 7.0 at 25°C.

Preparation: Dissolve 39 g in 1 liter of distilled or deionized water, autoclave, cool to 45–50°C, and dispense to tubes.

HEKTOEN ENTERIC (HE) AGAR

Acid fuchsin	0.1 g
Agar	14
Bile salts #3	9
Brom thymol blue	0.065
Ferric ammonium citrate	1.5
Lactose	12
Proteose peptone	12
Saccharose	12
Salicin	2
Sodium chloride	5
Sodium thiosulfate	5
Yeast extract	3

Final pH 7.5 at 25°C.

Preparation: Suspend 76 g in 1 liter of distilled or deionized water, heat to boiling, do NOT autoclave, cool to 45–50°C, and dispense to plates.

LACTOBACILLI MRS (DEMAN, ROGOSA, AND SHARPE) BROTH

Ammonium citrate	2 g
Beef extract	10
Disodium phosphate	2
Glucose	20
Magnesium sulfate	0.1
Manganese sulfate	0.05
Proteose peptone #3	10
Sorbitan monooleate complex	1
Sodium acetate	5
Yeast extract	5

Final pH 6.5 at 25°C.

Preparation: Suspend 55 g of MRS broth in 1 liter of distilled or deionized water, heat to boiling, and autoclave.

LACTOBACILLI MRS AGAR

Prepared from lactobacilli MRS broth with addition of 15 g/L agar. Dispense to plates.

LACTOBACILLI MRS SOFT AGAR

Prepared from lactobacilli MRS broth with addition of 7.5 g/L agar. Dispense to tubes.

LACTOSE BROTH (LB)

Beef extract	3 g
Lactose	5
Peptone	5

Final pH 6.9 at 25°C.

Preparation: Suspend 13 g of LB broth in 1 liter of distilled or deionized water, warm slightly, dispense in tubes, and autoclave.

LAURYL SULFATE TRYPTOSE (LST) BROTH

Lactose	5 g
Potassium phosphate, dibasic	2.75
Potassium phosphate, monobasic	2.75
Sodium chloride	5
Sodium lauryl sulfate	0.1
Tryptose	20

Final pH 6.8 at 25°C.

Preparation: Suspend 35.6 g of LST broth in 1 liter of distilled or deionized water, warm slightly, dispense to tubes containing inverted Durham tubes, and autoclave.

LYSINE IRON AGAR (LIA) SLANTS

Agar	15 g
Brom cresol purple	0.02
Ferric ammonium citrate	0.5
Glucose	1

L-Lysine hydrochloride	10
Peptone	5
Sodium thiosulfate	0.04
Yeast extract	3

Final pH 6.7 at 25°C.

Preparation: Suspend 34.5 g in 1 liter of distilled or deionized water, heat to boiling, dispense to tubes, autoclave, and cool medium in a position that will provide a short slant and a deep butt.

MANNITOL SALT AGAR (MSA)

Agar	15 g
Beef extract	1
D-Mannitol	10
Phenol red	0.025
Proteose peptone #3	10
Sodium chloride	5

Final pH 7.4 at 25°C.

Preparation: Suspend 41 g of the medium in 1 liter of distilled or deionized water, boil, dispense into tubes, autoclave, and dispense into plates.

MANNOSE (M) BROTH

Dipotassium phosphate	5 g
D-Mannose	2
Ferrous sulfate	0.04
Magnesium sulfate	0.8
Manganese chloride	0.14
Sodium chloride	5
Sodium citrate	5
Tryptone	12.5
Tween 80	0.75
Yeast extract	5

Final pH 7.0 at 25°C.

Preparation: Suspend 36.2 g in 1 liter of distilled or deionized water, heat to boiling, dispense into tubes, and autoclave.

MODIFIED *Escherichia coli* (EC) BROTH WITH NOVOBIOCIN (mEC+N)

Bile salt #3	1.12 g
Lactose	5
Potassium phosphate, dibasic	4
Potassium phosphate, monobasic	1.5

Sodium chloride	5
Tryptose	20
Novobiocin antibiotic supplement	10 ml

Final pH 6.9 at 25°C.

Preparation: Rehydrate novobiocin antimicrobic supplement with 10 ml of sterile distilled or deionized water, dissolve 36.6 g modified EC medium in 1 liter of distilled or deionized water and autoclave, cool to room temperature, add novobiocin antimicrobic supplement, and dispense aseptically to tubes.

Novobiocin antibiotic supplement
(per 10-ml vial)

Sodium novobiocin	20 mg

MODIFIED OXFORD (MOX) AGAR

Oxford medium base

Agar	2 g
Columbia blood agar base	39
Esculin	1
Ferric ammonium citrate	0.5
Lithium chloride	15

Final pH 7.2 at 25°C.

Preparation: Rehydrate modified Oxford antimicrobic supplement with 10 ml sterile distilled or deionized water, suspend 57.5 g of Oxford medium base in 1 liter of distilled or deionized water, heat to boiling, autoclave and cool to 45–50°C, add modified Oxford antimicrobic supplement, mix, and dispense to plates.

Modified Oxford antimicrobic supplement
(per 10-ml vial)

Colistin sulfate	10 mg
Moxalactam	20 mg

PALCAM MEDIUM BASE

Acriflavin hydrochloride	0.005 g
Agar	2
Columbia blood agar base	39
Esculin	1
Ferric ammonium citrate	0.5
Glucose	0.5
Lithium chloride	15
Mannitol	10
Phenol red	0.08

| Polymyxin B sulfate | 0.01 |
| PALCAM antimicrobic supplement | 2 ml |

Final pH 7.2 at 25°C.

Preparation: Rehydrate PALCAM antimicrobic supplement with 10 ml distilled or deionized water, suspend 68 g PALCAM medium base in 1 liter of distilled or deionized water, heat to boiling, autoclave and aseptically add 2 ml PALCAM antimicrobic supplement, mix, and dispense to plates.

PALCAM antimicrobic supplement
  (per 10-ml vial)

| Ceftazidime | 40 mg |

PEPTONE

Used as a 0.1% sterile solution to dilute food samples.

PHENOL RED SORBITOL BROTH

Beef extract	1 g
Phenol red	0.018
Proteose peptone #3	10
Sodium chloride	5
Sorbitol	5

Final pH 7.4 at 25°C.

Preparation: Suspend 21 g of phenol red sorbitol agar in 1 liter of distilled or deionized water, add 1 g of agar per liter of medium, heat to boiling, dispense to tubes containing inverted Durham tubes, and autoclave.

PLATE COUNT AGAR (PCA)

Agar	15 g
Glucose	1
Tryptone	5
Yeast extract	2.5

Final pH 7.0 at 25°C.

Preparation: Suspend 23.5 g in 1 liter of distilled or deionized water, heat to boiling, autoclave, and dispense to plates.

PCA WITH ANTIBIOTICS (OR ANTIBIOTIC PCA)

Same recipe for PCA with the addition, following autoclaving, of 2 ml of a 500-mg solution of chlortetracycline HCl and chloramphenicol per 100 ml of PCA.

POTATO DEXTROSE AGAR (PDA), ACIDIFIED

Agar	15 g
Glucose	20
Infusion from potato	200
10% Sterile tartaric acid	1.85 ml/100 ml

Final pH: 5.6 at 25°C.

Preparation: Suspend 39 g of PDA base in 1 liter of distilled or deionized water, heat to boiling, autoclave, and dispense to plates.

Pseudomonas ISOLATION AGAR (PIA)

Agar	13.6 g
Irgasan	0.025
Magnesium chloride	1.4
Peptone	20
Potassium sulfate	10

Final pH 7.0 at 25°C.

Preparation: Suspend 45 g of PIA base in 980 ml distilled or deionized water, add 20 ml of glycerol, heat to boiling, autoclave, and dispense to plates.

SELENITE CYSTINE (SC) BROTH

Disodium phosphate	10 g
Lactose	4
L-Cystine	0.01
Sodium acid selenite	4
Tryptone	5

Final pH 7.0 at 25°C.

Preparation: Suspend 23 g in 1 liter of distilled or deionized water, heat to boiling, dispense in tubes to a depth of 60 mm, and do NOT autoclave. Use immediately.

TETRATHIONATE (TT) BROTH

Bile salts	1 g
Calcium carbonate	10
Proteose peptone	5
Sodium thiosulfate	30
Iodine solution	2 ml/100 ml

Final pH 8.4 at 25°C.

Preparation: Rehydrate iodine solution with 20 ml distilled or deionized water, suspend 4.6 g in 100 ml distilled or deionized water, heat to boiling, do NOT autoclave, cool to 60°C, add iodine solution, and dispense into tubes. Use immediately.

Iodine solution
(per 20 ml distilled H_2O)

Iodine crystals	6 g
Potassium iodide	5 g

THIOGLYCOLLATE AGAR

Agar	20 g
Fluid thioglycollate medium	29.5

Final pH 7.1 at 25°C.

Preparation: Suspend ingredients in 1 liter of distilled or deionized water, heat to boil, dispense into tubes, autoclave, and use within a week.

Fluid thioglycollate medium

Agar	0.75 g
Casitone	15
Cystine	0.5
Glucose	5.5
Sodium chloride	2.5
Sodium thioglycollate	0.5
Resazurin	0.001
Yeast extract	5

TOLUIDINE BLUE-DNA AGAR

Agar	10 g
Calcium chloride	0.0011
Deoxyribonucleic acid (DNA)	0.3
Sodium chloride	10
Toluidine blue O	0.083
Tris(hydroxymethyl) aminomethane	6.1

Preparation: Dissolve tris(hydroxymethyl) aminomethane in 1 liter of distilled water. Adjust pH to 9.0. Add the remaining ingredients except toluidine blue O and heat to boiling to dissolve. Dissolve toluidine blue O in medium. Sterilization is not necessary if used immediately.

TRIPLE SUGAR IRON (TSI) AGAR SLANTS

Agar	12 g
Beef extract	3
Ferrous sulfate	0.2
Glucose	1
Lactose	10
Peptone	15
Phenol red	0.024
Proteose peptone	5
Sodium chloride	5
Sodium thiosulfate	0.3
Sucrose	10
Yeast extract	3

Final pH 7.4 at 25°C.

Preparation: Suspend 65 g TSI agar base in 1 liter of distilled or deionized water, heat to boiling, dispense into tubes, autoclave, and cool in position to provide short slant and deep butts.

TRYPTIC SOY AGAR (TSA)

Agar	15 g
Sodium chloride	5
Soytone (papaic digest of soybean meal)	5
Tryptone (pancreatic digest of casein)	15

Final pH 7.3 at 25°C.

Preparation: Suspend 40 g of TSA agar base in 1 liter of distilled or deionized water, heat to boiling, autoclave, and dispense to plates.

TRYPTIC SOY BROTH (TSB)

Dipotassium phosphate	2.5 g
Glucose	2.5
Sodium chloride	5
Soytone (papaic digest of soybean meal)	3
Tryptone (pancreatic digest of casein)	17

Final pH 7.3 at 25°C.

Preparation: Suspend 30 g of TSB in 1 liter of distilled or deionized water, dispense to tubes, and autoclave.

TRYPTONE GLUCOSE EXTRACT (TGE) AGAR

Agar	15 g
Beef extract	3
Glucose	1
Tryptone	5

Final pH 7.0 at 25°C.

Preparation: Suspend 24 g in 1 liter of distilled or deionized water, heat to boiling, autoclave, and dispense to plates.

UVM MODIFIED *Listeria* ENRICHMENT BROTH

Acriflavin hydrochloride	0.012 g
Beef extract	5
Esculin	1
Nalidixic acid	0.02
Potassium phosphate, monobasic	1.35
Sodium chloride	20
Sodium phosphate, dibasic	9.6
Tryptose	10
Yeast extract	5

Final pH 7.2 at 25°C.

Preparation: Suspend 52 g in 1 liter of distilled or deionized water, heat to boiling, dispense into tubes, and autoclave.

VIOLET RED BILE (VRB) AGAR

Agar	10 g
Bile salts #3	1.5
Crystal violet	0.002
Lactose	10
Neutral red	0.03
Peptone	7
Sodium chloride	5
Yeast extract	3

Final pH 7.4 at 25°C.

Preparation: Suspend 41.5 g in 1 liter of distilled or deionized water, heat to boiling (<2 min), do NOT autoclave, and dispense agar to plates.

XYLOSE LYSINE DESOXYCHOLATE (XLD) AGAR

Agar	15 g
Ferric ammonium citrate	0.8

Lactose	7.5
L-Lysine	5
Phenol red	0.08
Saccharose	7.5
Sodium chloride	5
Sodium desoxycholate	2.5
Sodium thiosulfate	6.8
Xylose	3.75
Yeast extract	3

Final pH 7.4 at 25°C.

Preparation: Suspend 57 g of XLD agar base in 1 liter of distilled or deionized water, heat to boiling, do NOT autoclave, and dispense to plates.

INDEX

Acinetobacter, 24
Activity arbitrary units, 235
Aeromonas, 24, 227
Alicyclobacillus, 24, 81, 83
Alternaria, 44, 46
Anaerobic incubation, 87, 88
Antibodies, 117
Antigens, 118
Ascomycota, 43, 44, 45, 47
Ascospore, 24, 43, 44
Aseptic technique, 5, 13, 15
Aspergillus, 24, 44, 46, 57, 58

Bacillus, 18, 24, 80–84, 113, 227
Bacteriocin, 224, 227
  activity, 235
  bioassay, 231, 233, 243, 246
  production, 226, 231
Baird–Parker agar, 124–125, 261
Basidiomycota, 43, 44
Basidiospore, 43, 44
Biochemical methods, 114, 115, 117, 141, 143–155, 171–187
Biopreservation, 226
Bioreactor, 231
Bismuth sulfite (BS) agar, 174, 176, 261
Botulism, 112, 113
Brain-heart infusion (BHI), 100, 125, 262
Brettanomyces, 47

Brilliant green lactose bile (BGLB), 63, 65, 262
Brochothrix, 24
Byssochlamys, 24, 44, 45

CAMP test, 139, 140, 155
Campylobacter, 24, 253–256
Candida, 24, 48
Capsular antigens, 168
Catalase test, 151, 154
Chytridiomycota, 43
Citrobacter, 24
Cladosporium, 44, 46
Clostridium, 24, 81, 83–84, 112, 226, 227
Coagulase, 122–124, 126, 134
Cold enrichment, 106, 139, 140
Coliform, 61–79, 175
Completed test, 62, 64
Confirmatory test, 62, 64
Conidiospore, 43
Corynebacterium, 224
Counting, 9–12, 18–19
Counting rules, 10–12, 27
Cryptococcus, 44, 48
Cryptosporidium, 24
Culture-based methods, 114, 116–117, 121, 141, 143–155, 173–187
Cyclospora, 24

Death phase, 229

Decimal dilution, 7, 30
Detection methods, 113
Deuteromycota, 43, 44, 46, 48
Dilution factor, 7, 9, 10, 234, 235, 255
DNA probe, 114, 118, 156, 167, 196, 198, 201–204

Enrichment, 114-116
Entamoeba, 24
Enterobacter, 61, 67, 78
Enterobacteriaceae, 167, 206, 212
Enterococcus, 78, 149, 151, 215, 216, 224, 227
Enumeration, 114, 117
Environmental sampling methods, 98, 102, 103
Enzyme immunoassay, 114, 118
Enzyme-linked immunosorbent assay (ELISA), 118, 167, 188, 206, 210
Eosin methylene blue agar, 63, 65, 263
Escherichia, 24, 61, 167
Escherichia coli, 15, 57, 58, 61, 78, 112, 206, 207, 211, 215, 216
Escherichia coli O157:H7, 111, 112, 206–222, 227
Eurotium, 44, 48
Exponential phase, 228

Febrile gastroenteritis, 139
Fermentor, 231
Flagellar antigen, 168, 207
Flavobacterium, 24
Food handling, 29–30
Food homogenization, 30
Fraser broth, 145, 263
Fusarium, 24, 44, 46

Gel electrophoresis, 141, 156–157, 162
Gel photographing, 158
Generation time, 229, 230, 258
Genetic methods, 114, 118–120, 138, 141, 156–164, 169, 173, 196–204
Gene-Trak, 196
Geotrichum, 24, 44, 46
Germinant, 86
Giardia, 24
β-Glucuronidase, 206, 220
Gram-negative broth, 196, 264
Gram staining, 20
Growth kinetics, 226, 231, 257–259
Growth phases, 228-229

Hanseniaspora, 24, 46

Hektoen enteric (HE) agar, 175, 176, 264
Hemolysis, 122, 138, 139, 140, 146
Hemolytic uremic syndrome, 207
Hepatitis A, 24
Heterolactic fermentation, 225
Homolactic fermentation, 225

Identification, 113-115, 141, 143, 173
Immunochromatographic assay, 210–211, 219
Immunological methods, 114, 115, 117–119, 171, 173, 188–195
Incubation, 7-8
Indole test, 211
Intoxication, 111
Isolation, 114, 115, 117, 141, 143, 173, 208

Klebsiella, 24, 60
Kluyveromyces, 24, 47

Lactic acid bacteria, 23, 224–226
Lactobacilli MRS, 237, 264–265
Lactobacillus, 24, 224–225
Lactococcus, 24, 224–225
Lactose broth, 176, 265
Lag phase, 228
Lancefield group N, 225
Lauryl sulfate tryptose broth, 63, 66, 265
Leuconostoc, 24, 224–225
Listeria, 8, 23, 97, 99, 101, 111, 112, 120, 138–166, 226–227
Listeriosis, 139
Lysine iron agar, 176, 177, 265–266

Mannitol salt agar (MSA), 126, 266
Mannose broth, 188, 266
Media preparation, 6, 52
4-Methylumbelliferyl-β-D-glucuronide (MUG), 206
Microbiological media, 5, 25, 117, 261
Micrococcus, 24, 227
Microscopic examination, 19-20, 31, 56–58, 114
Modified E. coli broth with novobiocin, 211, 266–267
Modified Oxford (MOX) agar, 145, 267
Monilia, 44, 46
Most probable number technique, 64–65, 71
Mucor, 24, 43, 44, 45

Neosartorya, 44, 45
Nisin, 226

Noninvasive infections, 111
Norwalk virus, 23
NOW kit, 209, 210, 219
Nucleic acid hybridization, 114, 118, 198

Oenococcus, 225

PALCAM agar, 145, 267–268
Pediocin, 226, 227
Pediococcus, 24, 224–225
Penicillium, 24, 43, 44, 46, 57, 58
Peptone water, 13
Personal safety, 1–2, 92, 111, 115, 126, 142, 165, 173, 212, 226
Petrifilm, 33, 51, 65
Phenol red sorbitol broth, 212, 268
Pichia, 44, 47
Plantaricin, 227
Plate count agar (PCA), 13, 14, 33, 99, 268
Plate count agar with antibiotics, 50, 52, 99, 268–269
Polymerase chain reaction (PCR), 114, 119–120, 138, 141, 156–164
Potato dextrose agar (PDA), 49, 50, 52, 269
Pour plating, 14, 17
Presumptive test, 61, 63
Propionibacterium, 227
Propionicin, 227
Pseudomonas, 23, 24, 97–107, 227, 229
Pseudomonas isolation agar (PIA), 101, 269

Rainbow agar, 212
Rhizopus, 24, 44, 45, 57, 58
Rhodotorula, 24, 47
Ribotyping, 113, 114
Rotavirus, 24

Saccharomyces, 24, 43, 47, 57, 58
Salmonella, 8, 23, 24, 97, 111, 113, 117, 118, 167–205
Salmonella-Tek, 188
Salmonellosis, 168
Selenite cystine broth, 176, 178, 269
Sensitivity, 115–116
Serotyping, 114, 115, 118
Serratia, 24
Shiga-toxin, 206
Shigella, 24, 112, 167
Somatic (O) antigen, 167, 207
Specific growth rate, 229, 230, 257, 259
Specificity, 115–116

Sporangiospores, 24, 43
Spore activation, 85
Spore cortex, 84, 85
Spore germination, 85
Spore outgrowth, 86
Spore straining, 94
Spore structure, 84
Spore-to-cell transition, 85
Spread plating, 14, 17
Staining mold, 57
Staphylococcal enterotoxin, 121, 127
Staphylococcal gastroenteritis, 112, 121
Staphylococcus, 24, 111–113, 121–165, 227, 238
Stationary phase, 229
Stomacher, 29
Streaking, 16
Streptococcus, 24, 224–225

Tetragenococcus, 225
Tetrathionate broth, 178, 269–270
Thamnidium, 44, 45
Thermophilin, 227
Thioglycollate agar, 88, 270
Toluidine blue DNA agar, 126, 270
Trichothecium, 44, 46
Triple sugar iron (TSI) agar, 176, 178, 271
Trypotose broth, 212
Tryptic soy agar, (TSA) 146, 271
Tryptic soy agar with blood, 146
Tryptic soy agar with yeast extract, 146
Tryptone glucose extract (TGE) agar, 87, 88, 272
Typhiod fever, 168
Typing, 113–115

UVM1 broth, 145, 272

Vagococcus, 225
Vibrio, 24, 112, 227
Violet red bile agar, 62, 66, 272

Wet mount, 19

Xylose lysine desoxycholate agar, 176, 179, 272–273

Yersinia, 24, 167, 227

Zygomycota, 43, 44, 45
Zygosaccharomyces, 24, 47
Zygospore, 43, 44